T0275942

SpringerBriefs in Environment, Security, Development and Peace

Volume 26

Series editor

Hans Günter Brauch, Mosbach, Germany

More information about this series at http://www.springer.com/series/10357
http://www.afes-press-books.de/html/SpringerBriefs_ESDP.htm
http://www.afes-press-books.de/html/SpringerBriefs_ESDP_26.htm

Liana Ricci

Reinterpreting Sub-Saharan Cities through the Concept of Adaptive Capacity

An Analysis of Autonomous Adaptation in Response to Environmental Changes in Peri-Urban Areas

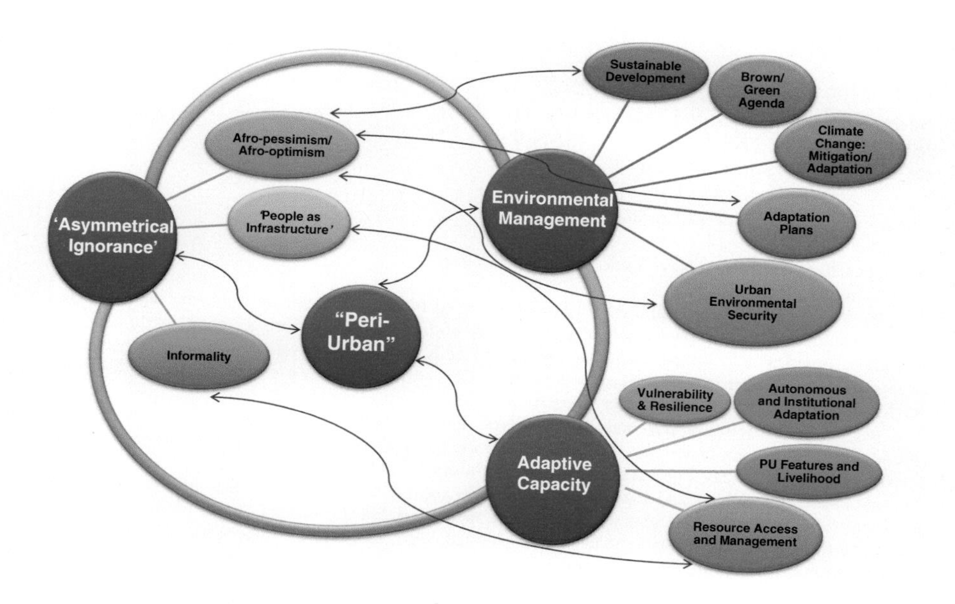

Liana Ricci
Department of Civil, Building
and Environmental Engineering
SAPIENZA University of Rome
Rome
Italy

Translation from Italian into English: Ashleigh Rose, Berlin (Germany)

Cover is based on the author's research model for which she owns the copyright

More on this book is at: <http://www.afes-press-books.de/html/SpringerBriefs_ESDP_26.htm>

ISSN 2193-3162 ISSN 2193-3170 (electronic)
SpringerBriefs in Environment, Security, Development and Peace
ISBN 978-3-319-27124-8 ISBN 978-3-319-27126-2 (eBook)
DOI 10.1007/978-3-319-27126-2

Library of Congress Control Number: 2015955856

Springer Cham Heidelberg New York Dordrecht London

Copyediting: PD Dr. Hans Günter Brauch, AFES-PRESS e.V., Mosbach, Germany

Printed on acid-free paper

Springer International Publishing AG Switzerland is part of Springer Science+Business Media (www.springer.com)

Acknowledgements

This work would not have been possible without the support of Prof. Silvia Macchi, Associate Professor of Urban and Regional Planning at Sapienza University of Rome (Italy), who greatly inspired and motivated me as the supervisor of my Ph.D. thesis on which this book is based. I am also deeply grateful to Prof. Paolo Colarossi for the essential input and support he provided throughout my Ph.D. programme, to the members of the supervisory committee, and to Prof. Gabriel Kassenga, Dr. Stephen Mbuligwe and engineer Nyamboge Chacha from Ardhi University (ARU), all of whom provided invaluable assistance and relevant input during the fieldwork conducted in Dar es Salaam (Tanzania). I also thank my colleagues in the Ph.D. programme for their generous sharing and discussions, and Carlo and Laura for their support in the final stage of the revision process.

I thank Loredana Cerbara and Marcella Prosperi at the Institute for Research on Population and Social Policies of the Italian National Research Council for their support during the data analysis. Many thanks are also due to Ashleigh Rose for the English translation.

In addition, I would like to thank Prof. Willard Kombe, Prof. John Lupala and Dr. Rubera A. Mato from Ardhi University, and Prof. Pius Zebhe Yanda, Dr. A. Majule, Dr. A. Mwakaje and Prof. Camillus J. Sawio from the IRA Institute of Dar es Salaam University (Tanzania) for their kind availability and input.

I am grateful also to Prof. David Simon of the Department of Geography at the Royal Holloway University of London, Dr. David Dodman at the International Institute for Environment and Development, and Dr. Julio D. Dávila of the Development Planning Unit at the University College of London.

I thank the officers from several Tanzanian Institutions: Mrs. Sarah Kyessi of the Tanzanian Ministry of Lands, Housing, and Human Settlements Development, Mr. Fredrick Mulinda of the National Environment Management Council (NEMC), Mrs. Ester Kibona from Environmental Protection Management Services (EPMS), and Mr. Praygod Mawalla of the Belgian Technical Cooperation for providing information beneficial to the research.

For their kind support during the fieldwork, thanks are also due to the technical officers of Kinondoni Municipality in Dar es Salaam, Mrs. Lucy B. Kimoi, Mrs. Mary Komba and Eng. E. Mwampashi, and to the Dar es Salaam ward leaders and community leaders, Mr. Said Kikwi from Bunju A subward, Mr. Ndagile from Makongo subward, Mr. Kambi from Kawe ward, Mrs. Idda Temba from Kawe ward, Mr. Pendo Fridy from Kawe ward, and Mr. Seif and Mr. Ryceson Kajiri from Msasani ward.

Special thanks go to editors Hans Günter Branch and Johanna Schwarz and to the anonymous reviewers for their insightful assistance and very useful comments, which led to the publication of this revised Ph.D. thesis.

Rome, Italy
June 2015

Liana Ricci

Contents

List of Figures

Chapter 1
Introduction

Abstract The motivation for this research is described with reference to the author's previous experience in sub-Saharan Africa and the need to further explore environmental issues in African cities. Acknowledging the gaps between urban planners' interpretive models and the daily activities of the people who create the city, the research investigates whether the concept of adaptive capacity can facilitate an interpretation of the sub-Saharan (and contemporary) city that is able to overcome the dichotomies, categories and partial approaches of prevailing urban geography. Fieldwork in peri-urban areas of Dar es Salaam defined the main factors of adaptive capacity: type and magnitude of local environmental impacts of climate change; rural–urban dynamics and relations; local autonomous adaptive capacity; institutional capacity in environmental management and urban development planning. This research generated knowledge on the dynamics of development and environmental management, as well as methodological insights regarding the limits and strengths of the research approach applied.

Keywords African cities · Urban development · Environmental change · Sub-Saharan · Urban studies · Research question

1.1 Context and Motivation

Reflecting a posteriori on the reasons I chose the urbanization process in the cities of sub-Saharan Africa as topic of study, I think my interest in the subject was sparked by research I had previously conducted in Lusaka and the Zambian Copperbelt cities while preparing my Bachelor thesis. That interest continued to develop during my Master's and Doctorate, during which I studied land use and environmental planning theories and instruments. The Copperbelt cities, known by historical anthropologists and sociologists as the sites of the most difficult and deleterious urbanization during the colonial period, have since experienced the even worse impacts of subsequent exploitation by the mining industry. In 2003, the

L. Ricci, *Reinterpreting Sub-Saharan Cities through the Concept of Adaptive Capacity*, SpringerBriefs in Environment, Security, Development and Peace 26, DOI 10.1007/978-3-319-27126-2_1

'urban' reality that I had the opportunity to observe in five different cities, achieved ad hoc by mining corporations, was no better (and perhaps worse) than that described in studies of the colonial period: spatial segregation and social discrimination, environmental exploitation and destruction. Despite payment of environmental fines by mining corporations, phenomena such as the heightened mortality attributable to water and soil pollution by copper processing plants, and high workplace mortality were ignored in the name of local economic development.

As a result, questions began to arise regarding the relationship between urban development and environmental transformation, the 'power' of planning, and the close relationship between people's lives and the management of natural resources. These questions are relevant not only in Africa, but also in Western cities, which some historians view as having fairly risky implications not much different from their southern counterparts.

Given this point of departure, the reasons that motivated me to undertake this course of research are, on one hand, the conviction that African cities are not something conceived by 'others' that exist "in the perspective of becoming the other" (Cassano 1996: viii), the future of which is foreseen as "an eternally incomplete and failed pursuit" (Cassano 1996: viii). Rather, those cities have a cultural autonomy and specific modality for the production of space. Perhaps the greatest challenge for this kind of research project is discovering how this specificity can be brought to light without falling into the trap of overemphasizing the otherness and the exotic and seductive forms of the cities of the South.

In addition, one must confront the limits and the difficulties of a research process that uses instruments and 'pre-structures' which are the product of a Western (Italian) background, while focusing on social and environmental phenomena and practices that maintain the richness of a 'non-modern' African culture, despite their contact with Western culture since the colonial period.

The hybrids, conflicts, and differentiations that have sprung from the blending of many points of view constitute, in my opinion, the greatest asset of this study, and have allowed it to encompass diverse perspectives, horizons, languages, and people. As such, the study of cultural contamination as a resource in the research process is reflected in the continued attempt to adopt an approach that integrates theoretical and practical dimensions, while bearing in mind the limited nature of the spatial-temporal and cultural context on which the knowledge process is based.

Many cities of sub-Saharan Africa are experiencing rapid demographic growth and parallel expansion in urbanized areas. The global urban population surpassed the rural population for the first time in 2009, mostly as a result of growth in these cities.

This growth is taking place predominantly in areas that have been defined by a vast body of literature as peri-urban, areas with hybrid rural urban characteristics that develop through informal, unplanned modalities, and where the relationship with (in some cases the dependence on) natural resources has a crucial role in people's living strategies.

In most urban studies, informal and unregulated development is the structural cause of the environmental, social, and economic criticality that affects African cities, particularly peri-urban areas, and of ineffective planning that is unable to govern such rapid and 'unruly' transformation of space.

Alternatively, some authors attribute the ineffectiveness of planning to inadequate interpretive approaches. This includes the persistence, in the regulatory sphere, of approaches pervaded by 'asymmetrical ignorance', and of the hegemonic and dominant position of Western scientific production, which often assumes or produces universalization strategies and practices of exclusion from knowledge production. This debate has become more heated in recent years in the face of aggravated environmental 'crises' linked to *Global Environmental Change* (GEC), highlighting the extreme vulnerability of these cities and the urgent need to formulate adequate approaches to interpretation and planning.

The study assumes, in accordance with the second above-cited position developed predominantly in post-colonial studies, that the imposition of Western interpretive models could exacerbate the effects of environmental transformation already underway. It calls into question the approaches that conceive of sub-Saharan peri-urban areas in 'negative' terms, considering them incomplete or somehow 'lacking', and requiring support in the inevitable transition process towards an urban and more secure, regulated, and formal condition. Instead, the study embraces the approaches that share a 'positive' vision of those areas, focusing on what exists and is already happening in them, on the practices of production of informal space, and the ephemeral boundary that both separates them from and interweaves them with formal areas.

At this point in the research process, the need arises for a closer investigation of the environmental management practices that allow people in peri-urban areas to organize their own living strategies and to face the challenges of environmental change, created by local climate change or by its interaction with urbanization processes.

The concept of 'adaptive capacity', developed in the climate change debate, refers to the capacity of an 'urban system' to successfully address the transformations underway by reducing the vulnerability to those changes of the people (or sub-systems) within it. This capacity is the product of measures taken by institutions and autonomous strategies undertaken by individuals in a manner that would be defined as informal. These strategies and measures coexist in a complex relationship that is not always synergetic and sometimes even conflicting. That relationship has strong political characteristics and is of great interest as regards planning.

From this perspective, an investigation of the factors determining capacity for autonomous adaptation to environmental changes in the peri-urban areas of Dar es Salaam becomes the instrument for bringing to the fore the informal 'platforms of action' around which life is constructed in peri-urban areas, and for underlining the importance of such platforms, and of spaces with hybrid rural–urban characteristics, in terms of access to and management of natural resources. Moreover, this allows one to highlight the contradictions in the interpretive models that planners use to understand the urban sub-Saharan reality, as well as the gap between those models and the daily activities of the people who live in and create the city.

1.2 The Research Question

The primary goal of this study is to bridge that gap by identifying the cognitive and methodological elements that are considered indispensable in any planning process for intervention in contexts like the one analysed. The main question around which the study is developed is therefore the following: Can viewing the sub-Saharan city through the concept of adaptive capacity contribute to construction of an interpretation of the contemporary city that overcomes the dichotomies, categories, and partial approaches of prevailing urban geography?

To try to answer this question has first meant identifying the assumptions upon which the principle interpretive approaches for peri-urban areas of sub-Saharan cities are based, in order to then define a study area within Dar es Salaam. These assumptions include, on one hand, the presuppositions with which peri-urban areas are investigated (widespread and homogeneous poverty, subsistence economies based exclusively on agriculture, limited access to resources and services, etc.) and above all 'planned' or requalified; on the other hand, the presuppositions with which the vulnerability of those same areas, and therefore their adaptive capacity, is evaluated (inability to observe and manage environmental change, heightened exposure to transformations due to the absence of infrastructure and services, etc.). Given these two closely connected and interdependent groups of assumptions and the need to verify their applicability in the context of Dar es Salaam, two groups of inquiry were developed in order to respond to the research question.

1.3 Specific Objectives

Both the characterization of peri-urban areas (family, geographic and environmental characteristics, etc.), and the factors that constitute adaptive capacity in those areas, were investigated for the purpose of identifying reciprocal interdependencies and the limits or contradictions that those interdependencies provoke with respect to prevailing interpretive approaches.

The information on peri-urban areas constitutes the basis for defining the principal factors of adaptive capacity in the peri-urban context, factors that fall within four main areas: type and magnitude of local environmental impacts of climate change; rural–urban dynamics and relations; local autonomous adaptive capacity; institutional capacity in environmental management; and urban development planning.

1.4 Approach and Method

The study followed a recursive path in which successive analyses of the literature were integrated with observations and field studies. The literature analysis collected contributions to the questions outlined above from the leading schools of Western planning, but it favoured local literature from Tanzania and elsewhere in Africa, including ethnographic, anthropologic, social, and economic studies of sub-Saharan cities. Primary reference was made to the scientific productions of the *African Centre for Cities* (ACC) of the University of Cape Town, and of Ardhi University in Dar es Salaam, where some of the research was conducted.

Field research entailed administration of questionnaires to households residing in selected peri-urban areas north of Dar es Salaam, interviews with officials and researchers, and field observations.

The data collected through the questionnaires was not analysed using a quantitative approach, given the limited number of questionnaires. That data was the starting point for a critical review of the theoretical inputs obtained from the literature review, and was used to interpret direct observations made in the field.

1.5 Structure of the Book

This book is organized into six chapters, including this introduction. The second chapter outlines the theoretical and methodological framework that guided the research process, and lays out the arguments and reflections that led to the formulation of the research question. Inadequate interpretive and planning approaches are defined as being caused by ‘asymmetrical ignorance’, and the basic content of alternative approaches is identified, in particular, the ‘people as infrastructure’ approach. Finally, the knowledge project that was constructed using that approach, and the instruments and method with which the case study was conducted, are presented.

The third chapter reviews the key historical and political points in the urbanization process in sub-Saharan Africa. First, the difficult ‘modernization’ process promoted from time to time, first by colonial policies and later by development aid, is described. The distinct characteristics of sub-Saharan cities are presented, specifically their hybrid rural–urban features, their ‘informal’ development modality, and their relationship with environmental transformation. In the second part of the chapter, the case of Dar es Salaam is introduced as a representative example of the cities of sub-Saharan Africa, and part of the evidence obtained from the questionnaires or extracted from the literature is offered. The city is analysed with the evolution of peri-urban areas as the common thread, and the challenges faced by planning today are discussed, as well as some of the constraints of the policies and planning approaches adopted in Dar es Salaam and sub-Saharan Africa generally.

The fourth chapter, drawing from the challenges that emerged in the previous section, concentrates on the environmental question and on approaches to environmental management and urban planning that have become skewed over time in order to resolve the African 'environmental crisis' in the name of sustainable development and/or poverty reduction. These approaches have been brought to the fore by the global debate on Climate Change of the last few years. The effects and the extent of environmental transformations currently underway in sub-Saharan Africa and in Dar es Salaam are discussed, as well as the manner in which the two global strategies of Mitigation and Adaptation to Climate Change orient policy and planning for urban development at both the local and global levels. From that discussion emerges the crucial role of the Adaptation strategy in planning processes in African cities, which leads to a reconsideration of the impact of strategies which address 'securitization' of the city, rather than acceleration of the rural–urban transition, as a 'solution' for reducing social vulnerability. Such strategies raise questions that are not new to the planning debate, and draw attention once again to the role that people must play therein.

The fifth chapter focuses on the role of the autonomous practices for environmental management and adaptation to environmental transformations, adopted by people settled in peri-urban areas. This is the point of departure for the formulation and implementation of local adaptation action plans in the urban domain. The key factors in assessing peri-urban adaptive capacity are identified, and the results of the investigation carried out in Dar es Salaam are presented. Modalities of access to and management of resources, perception and observation of environmental transformation and current or prospective autonomous strategies for confronting it are discussed (in terms of the analysis of the questionnaires). Lastly, the relationship between the characteristics of the households interviewed (including geographical–environmental) and adaptation strategies identified is outlined in order to identify the specific interdependencies, the delicate equilibria, the limits, and the opportunities that derive from those interdependencies, and to understand how they may influence measures undertaken by institutions for adaptation to environmental transformation.

The conclusions (Chap. 6) of the research project are presented in the sixth chapter. In the first section, and in light of what emerged from the analysis of Dar es Salaam, the assumptions of the prevailing interpretive approaches in the peri-urban sphere are discussed, and several research questions are reformulated. The discussion is organized around several key issues according to which a new interpretation of peri-urban areas and sub-Saharan cities is constructed. These issues are:

- Relationship with natural resources
- Socio-economic and cultural homogeneity
- Environmental management and adaptive capacity
- 'People as infrastructure'
- The 'ideal of life' (urban?)
- Dynamism in the form of use and access to resources
- Rural–urban interdependence and bidirectional migration.

Lastly, two concluding considerations are offered on the limits or 'traps' of the research on peri-urban planning, including but not limited to sub-Saharan Arica. One of the first limits is related to the normative interpretive approaches that are aimed at transition from the peri-urban to urban state, and therefore provision of a 'modern' infrastructure conceived as steel and cement. These limits consist of the risk of formalizing, paralyzing or constraining social relations and 'platforms of action' within a structure that is incapable of responding to the needs of the people who live in peri-urban areas. A second limit is related to research, and the risk that an approach centred on the capacity of African citizens to act in an autonomous, informal, and effective way could prove to be a trap of self-exploitation and poverty that precludes development alternatives. Finally, these reflections are reassessed as a contribution to the renewal of interpretive approaches in the cities of the Global North, where the phenomena of rural–urban and informal hybridization are widespread and the impacts of environmental transformation linked to climate change are increasingly evident.

Figure 1.1 synthesizes the linkages between the guiding concepts addressed in the following chapters. Peri-urban areas are the specific topic of research, and are investigated using paradigms and concepts that are oriented around the theory of Asymmetrical Ignorance. This concept helps to develop a contextual description of the peri-urban areas in Dar es Salaam with a focus on the positive aspects of informal production of space and the role of human agency in shaping informal infrastructure (people as infrastructure). In particular, this perspective discusses the environmental management actions and approaches related mainly to climate change (Urban Environmental Security, Adaptation, plans, etc.) adopted in sub-Saharan cities, both

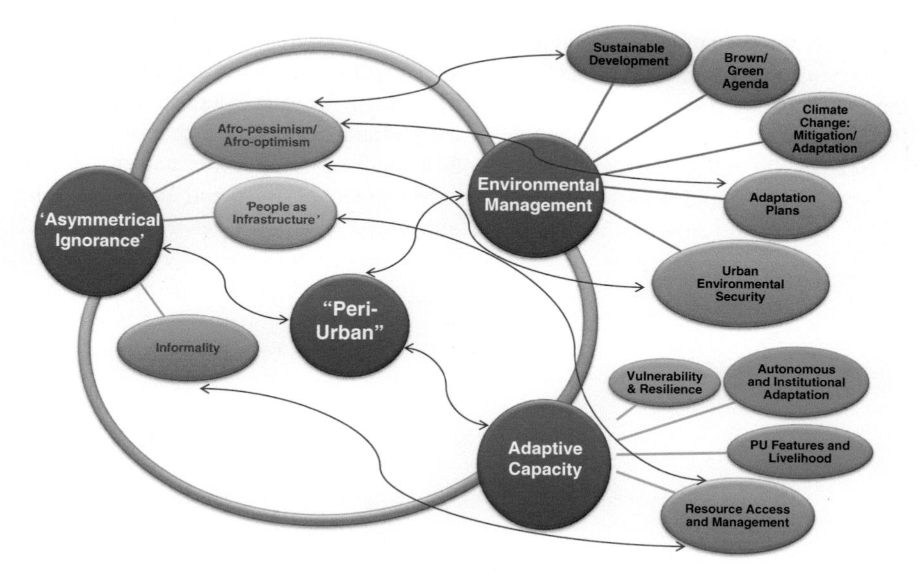

Fig. 1.1 Research conceptual framework. *Source* The author

those developed by institutions or development agencies and those developed by peri-urban residents. The concept of adaptive capacity enriches the discussion by linking post-colonial theories and analyses of agency to environmental management actions that address climate change adaptation, challenges, and opportunities. In particular, institutional (plans, strategies, policies, etc.) and autonomous (modalities of accessing and managing resources, migration, etc.) environmental management and climate change adaptation action/initiatives are influenced by the characteristics and conceptualization of peri-urban areas. Developing the research around the three main pillars of asymmetrical ignorance, environmental management, and adaptive capacity with a focus on peri-urban areas, this manuscript highlights the connections and interdependencies among them. The interconnections between the conceptualization and bio-physical (environmental) characteristics of peri-urban areas and residents' modalities of accessing and managing resources, building their livelihoods, and adapting to environmental changes is crucial to understanding the key elements of adaptive capacity.

1.6 Results

This research has made two main contributions. One consists of the knowledge produced on the dynamics of development and environmental management in peri-urban areas of the cities in sub-Saharan Africa, particularly Dar es Salaam. The second type of contribution is of a methodological nature, and consists in the totality of the limits and strengths of the process of investigating adaptive capacity and environmental management, tested on the city of Dar es Salaam, which may be readapted for the analysis of similar contexts. The investigation, oriented towards identifying what exists and what is occurring in peri-urban areas rather than what they are lacking in order to become 'urban', is configured as an ad hoc analytical framework for areas that share certain primary characteristics with peri-urban Dar es Salaam, including the presence of both urban and rural forms and activities, dynamism and rapid population growth, environmental stress and heavy dependence on natural resources, and predominantly informal modalities of production.

Chapter 2
Cities of Sub-Saharan Africa: Failed or Ordinary Cities?

Abstract This chapter outlines the theoretical and methodological framework that guided the research process. It presents a critical review of the main theoretical references, including post-colonial, informality, and vulnerability studies, and lays out the arguments and reflections that led to the formulation of the research question. Inadequate interpretive and planning approaches are defined as being caused by 'asymmetrical ignorance', and the basic content of alternative approaches is identified, particularly the 'people as infrastructure' approach. In order to explore the limits of asymmetrical ignorance, the study focuses on the interaction between urban development and environmental change in peri-urban areas of the sub-Saharan city by investigating households' environmental management practices. This knowledge is essential not only to shed light on the development and environmental management dynamics in peri-urban areas and the interdependence thereof with urban areas, but also to define the necessary conditions for effective adaptation to environmental change.

Keywords African urbanism · Informality · Post-colonialism · Urban bias · Rural–urban · Vulnerability · Agency · Black urbanism · Dar es Salaam

2.1 Urban Development and Planning: An Overwhelming Distance

There are two main reasons for challenging the generalizations made about contemporary African cities. First, because it allows one to move away from preconceived normative ideas about African cities that are based on 'urban planning' theories and practices developed in other parts of the world, particularly regions with global economic centrality, and towards an approach that accepts urban reality from a variety of different perspectives. Many studies, inspired mainly by post-colonial theories, have sought to overcome the categories and stereotypes of Western urban planning when examining African cities (Roy 2009; Freund 2007;

L. Ricci, *Reinterpreting Sub-Saharan Cities through the Concept of Adaptive Capacity*, SpringerBriefs in Environment, Security, Development and Peace 26, DOI 10.1007/978-3-319-27126-2_2

Robinson 2006; Simone 2004; Pieterse 2009; Murray/Myers 2006; Gandy 2005; Parnell/Pieterse 2014). Such studies represent an exciting moment in African urban thought, which is opening up to new interpretations and representations of the city. Some of these ideas concentrate on the myriad creative modes through which African 'urbanity' capitalizes on its own environments, and explore the difficulties and freedoms generated by life in African cities. These explorations are based on the idea that the residents of African cities can invent and develop creative strategies in order to mould their urban environment into flexible and appropriate spaces.

The second objective of this investigation of African cities (see Chaps. 3–6) is to build on the aforementioned studies in order to achieve a deeper understanding of the relationship between the city and natural resources, and between development of African cities, with all their unique characteristics, and the environmental changes currently underway. These objectives are reached through the case analysis of Dar es Salaam, specifically its peri-urban areas, based on fieldwork, desk research and discussion with other international scholars working on the topics of peri-urban areas, climate change adaptation, and environmental management in the cities of least developed countries. The elements of dynamism, complexity and diversity in contemporary African cities come to the fore as light is shed on the relationships that the people living in those cities have with natural resources, and on how urbanization conditions and is conditioned by such relationships, and by development policies and strategies at the global level.

2.1.1 Cultural Bias: From Post-colonial Studies to 'Asymmetrical Ignorance'

The human experience in Africa constantly arises in debates in the field of urban studies, as well as anthropology, sociology and economics as an experience that cannot be understood through an exclusively negative interpretation.

Post-colonial studies need an alternative to the dominant interpretation in which Africa is never considered a place that possesses characteristics of human nature, or when it is, such characteristics are given less value, considered less important, and of inferior quality. Through this elementary and primitive lens, Africa is rendered unfinished, lacking in something, incomplete. Mbembe (2001: 8–14), in his work *On the Postcolony*, identifies two keys to understanding the debate over Africa as a foreign and exotic land. The first refers to the idea of strange and monstrous discoveries made by abandoning familiar paradigms; Africa is to be understood for what it is, an entity with its own raw, brutal and savage characteristics. The second key is 'intimacy', according to which the African possesses a self-referential structure that renders him human, but belonging at the same time to a world that we cannot access, in which we can intervene to educate the African about our more human way of life (Mbembe 2001). It is on these bases that, according to Mbembe, Africa is constructed as an object of experimentation.[1]

Gathering a series of observations on how Africa is interpreted and represented, Mbembe reaches a tough conclusion that echoes the principles of old colonialism. He emphasizes how Africa is the image of foreignness *par excellence* in both daily life and in academia, and represents a universe at the margins of the Earth, where reason is quashed. He asserts that the inaccessibility of the obscure African universe is caused by the absence of an autonomous discourse on Africa: 'In the very principle of its constitution, in its language, and in its finalities, narrative about Africa is always pretext for a comment about something else, some other place, some other people. More precisely, Africa is the mediation that enables the West to accede to its own subconscious and give a public account of its subjectivity' (Mbembe 2001: 3). Such a scathing judgment is justified by the permanence of the prejudiced belief, still alive today, that African social groups belong in the category of simple and traditional societies.[2] From this prejudice derives the image of Africa as a *farmakos*, or the Western obsession with what is missing, with absence, non-being, identity, difference and negativity; in other words, with nothing. According to Mbembe, this image goes beyond the problem of Western thought as the *other with respect to the other*, as defined by de Certeau, or as the opposition between truth and error, reason and madness, as maintained by Foucault and Mouralis. Rather, it is a 'principle of language and classificatory systems in which to differ from something or somebody is not simply *not to be like* (in the sense of being non-identical or being-other); it is also *not to be at all* (nonbeing). More, it is *being nothing* (nothingness)' (Mbembe 2001: 4). It is in these terms that Africa is attributed with a particular unreality, the image of nothing, the abolished and the non-existent.

In debating how to highlight aspects of political thought and the political, social, and cultural reality of contemporary Africa in terms of their intrinsic value, or for comparative study with other societies, political science and economics still maintain the prejudices outlined above. Other disciplines, inspired by Foucauldian and neo-Gramscian or poststructuralist paradigms, such as historiography, anthropology, and feminist critique have concentrated, according to Mbembe, on the single problem of how *hybrid, fluid* and *negotiated* identities are invented (Mbembe 2001).

---

[1]An understanding of the various reasons for this state of affairs raises a series of questions about the condition and experience of the other, and the diversity with which the Western philosophical tradition has always clashed. In the relationship with Africa, the concept of 'absolute otherness' is recognized as a polemic and sometimes extreme argument, used in the West in order to affirm its own difference from the rest of the world. In many aspects Africa therefore constitutes a metaphor through which "the West brings into play the origins of its norms, constructing an image of itself and integrating it into the sum of indicators that reaffirm what it imagines to be its own identity" (Mbembe 2001).

[2]Mbembe refers to three main elements in order to characterize traditional societies: facticity and arbitrariness, a strong connection to the tradition of magic, and the symbolic and the prevalence of the person over the individual (Mbembe 2001: 11–12).

These disciplines, with the pretext of avoiding single-factorial explanations of domination, have reduced the complex phenomena of the state and power to 'discourses' and 'representations', while forgetting the material aspect of such discourses and representations. The rediscovery of the subaltern subject and the accent placed on its inventiveness have been transformed into perennial invocation of the notions of 'hegemony', 'moral economy', 'agency', and 'resistance' (Mbembe 2001). Mbembe's criticism of such disciplines is that nearly all authors, embracing the Marxist tradition, have continued to operate as though economic and material conditions of existence had an automatic reflex and find direct expression in a subject's consciousness; to explain the tension between structural determinants and individual actions, they fall into the trap of functionalism. Basing themselves on questionable dichotomies, these authors maintain that everything becomes clear as soon as one is able to demonstrate that the subjects of the action, subjugated (dominated) by power and the law (the colonized, women, farmers, labourers), possess rich and complex consciences that are capable of opposing their state of oppression; power is constantly contested and reconfigured until its own targets reappropriate it (Mbembe 2001).

Even after Marxism was no longer used as an analytic tool and pan-comprehensive project, and after dependence[3] theories were abandoned, a "false dichotomy between the objectivity of structures and the subjectivity of representations" continued to persist. This distinction allowed all that was cultural and symbolic to be relegated to one side, while everything economic and material (Mbembe 2001) was placed on the other. According to Mbembe, this constituted a rejection of the philosophical perspective that negated any reflection on African society and deprived it of legitimacy. Nevertheless, an instrumental paradigm that is too reductionist to shed light on the fundamental questions related to the social reality of Africa continues to dominate in every field.

Mbembe's position is based on two observations. One maintains that social reality in Africa is formed by a multiplicity of practices, products and objectives, and not merely discursive and linguistic practices, but also doing, seeing, listening, tasting, smelling, and touching. For all those who participate in the production of the 'African self', these practices represent 'significant human expressions' that render the African subject similar to every other human being who is engaged in 'significant acts' that nevertheless do not have the same meaning for another person (Mbembe 2001). The second observation is that the African subject does not exist independently of the actions that create social reality, separate from the process through which those practices have intrinsic meaning.

From the perspective of urban studies, Mbembe's text is imbued with inputs from the political geography of post-colonial power, which are abstractly theoretical and provocative, but are not particularly comprehensible in spatial terms

[3]According to the perspective of dependence theories, rapid urban growth, the commercialization of peri-urban activities, and the land market are considered destructive for the livelihoods of households and institutions (Mbiba/Huchzermeyer 2002: 125).

(Myers 2011: 45). The question of whether the colonial period has truly been overcome remains unanswered in African studies. A vast branch of African urban studies[4] (Beall/Fox 2009; Freund 2007; Mbembe/Nuttall 2004; Pieterse 2010; Murray/Myers 2006; Myers 2011; Simone 2004, 2010; Simone/Abouhani 2005; Roy 2005, 2009; Roy/AlSayyad 2004, Bryceson/Potts 2005) has recently developed around this basic question and on the common lens of development through which African cities are studied (Robinson 2006: x) and 'measured' with Western indicators (Myers 2011: 1). The hypothesis constructed by Jennifer Robinson is that there is a lack of knowledge on African cities primarily as a result of what Chakrabarty defines "asymmetrical ignorance" (Robinson 2003: 275), which is blinding and misleading.

2.1.1.1 Negative Interpretive Approaches

Although the urbanization process in sub-Saharan Africa has produced a series of interpretations among researchers, in the academic literature on African cities there is a general consensus that the accumulation of 'worrying' characteristics, such as unregulated growth, the scarcity of gainful employment opportunities in the formal economy, serious environmental decline, the lack of sufficient housing at accessible prices, the lack or inadequacy of infrastructure, the absence of basic social services, impoverishment, criminality, inadequate management of the city, and the increase in inequality, lead to a permanent condition of urban crisis (Rakodi 1997; Tostensen et al. 2001).

There is no one widely shared opinion regarding the causes of the 'urban crisis' underway, or on how to change the situation.

While some attribute the urban crisis mainly to rapid population growth (the demographic explosion) and to adverse economic conditions, others attribute it to corruption, poor management, or the failure of municipalities to provide the institutional and juridical supports necessary to stimulate entrepreneurial growth and development (Tostensen et al. 2001: 7, 10–11).

There are also notable differences among models of urban growth and development. In fact, most of the scientific research on African cities tends to ignore their differences and historical specificity, concentrating instead on common characteristics. In general, contemporary representations and debates have produced mechanistic (and simplistic) images of spatial incoherence, overcrowding, impoverishment, unemployment, decline, negligence, organized crime, daily violence, inter-ethnic conflicts, civil disorder, environmental degradation, pollution, rebellious behaviour, and juvenile delinquency (Murray/Myers 2006: 1).

[4]Many theorists, whose own roots are in post-structuralism and who choose to construct their work on African theories and practices, often emphasize the informal, the invisible or the new geographies of connection, movement, fluidity, flexibility, and contingencies as relevant in creating the urban areas of Africa (Myers 2011: 139).

This near-obsession as regards 'urban pathologies' and continuous failures—to the total exclusion of nearly everything else—reduces city life in Africa to a dystopic nightmare, where the "eschatological evocation of urban apocalypse" (Gandy 2005: 38) feeds the unilateral perception of such 'Afro-pessimists' (Murray/Myers 2006: 2), who suggest that the cities of Africa, and Africa in general, are so irredeemably chaotic and disorganized that they are far from their own 'redemption'.

Murray and Myers, drawing into the field of urban studies several aspects that have already been evinced by Mbembe, affirm that this disproportionate attention to the 'uncontrolled' or 'chaotic' urbanization process in Africa coincides with a kind of unreflecting sensationalism that has indelibly marked Africa as the 'Dark Continent', where 'tradition' and custom obscure rational-legal authority, where primordial 'tribal' devotions are rooted and permanent, where popular beliefs triumph over good sense, and the neo-patrimonial leadership works hand in hand with corruption, favouritism and widespread mismanagement of urban affairs (Abdoul 2005; Enwezor et al. 2002; Tostensen et al. 2001 in Murray/Myers 2006).

In the search for the 'meaning' of African cities, some look at the decay of physical infrastructure, the distribution of mechanisms for promoting social collaboration and the absence of the institutional framework necessary for social cohesion, and conclude that the cities in Africa simply 'don't work'. Others have attached themselves to the notion of 'African exceptionality', the idea that urban development works everywhere except in Africa, because the cities are intrinsically different and uniquely incapable (Roe 1999, cited in Murray/Myers 2006: 6).

These approaches concretize what can be defined as the *negative interpretive approach*, which has dominated much of the academic literature on urbanization in Africa. In fact, much of that literature implicitly or explicitly begins with normative prescriptions for how cities should ideally function. Framed in this way, urban sprawl in the African metropolis generally appears to be an exemplary expression of failed, distorted or impeded urban planning, in which the requisite bases and attributes of true urbanity that indicate the urbanization process elsewhere are missing.

2.1.1.2 Positive Interpretive Approaches

In contrast with the approach outlined above, a more fruitful *positive interpretive approach* has also been developed, which begins with the premise that the cities in Africa are, as Abdoumaliq Simone asserts in his *For the City Yet to Come: Changing African Life in Four Cities* (2004: 1–2), 'works in progress' moved forward by the inventiveness of ordinary people and held together by inertia and slow adaptation to changeable circumstances.

This new emphasis on temporariness allows one to conceive of African cities as places in which the possibility of becoming urban is not predetermined (Murray/Myers 2006: 6), in which enormous amounts of creative energy have been ignored and damaged (Simone 2004: 2), places that contain the seeds of new types of urban forms and solutions (Freund 2007: 171).[5]

Moving beyond a diagnostic mental attitude based on normative injunctions is a prerequisite to discovering and appreciating the historical specificity of African cities and how they simultaneously construct themselves and are constructed (Simone 2004: 15–16).

The proposal of Jennifer Robinson seems to exceed (overcome) the extreme faith and mistrust in African cities, maintaining instead an "ordinary city" perspective (Robinson 2006) according to which the city seems to resist a priori analytical classifications and hierarchies. This allows one to view the city as a historically specific entity, rather than an incarnation of abstract models (Amin/Graham 1997; Robinson 2006). Broadening the urban studies field of inquiry in this way provides a basis for deeper understanding of the nuances of diverse urban experiences and the complexity of urbanization in different parts of the world (Murray/Myers 2006: 10).

Simone (2010) asserts there is a need to begin with a discussion of cityness, from movements, to experiments, to 'peripheral' experiences of the 'black' city as a stimulus for overhauling the state of the periphery. He develops his argument by indicating five manifestations of peripheral life.

Peri-urban areas are one of these manifestataions and, like the rural–urban interface, they have been the subject of much of the debate on the 'periphery'. Analysis of such areas allows for the identification of cognitive elements that are useful in overcoming preconceived interpretations, since they represent a rich area in which rural–urban interdependence at the local and regional level and the connected ecological, social and economic relations are more evident (Simone 2010: 51–59).

Similarly, Pieterse (2008: 2–3) seeks to draw connections between analyses that address the daily practices in most urban areas. According to that author, it is in informal settlements that post-colonial cityness is most likely to emerge. Such informal settlements represent an area of opportunity and autonomy in various spaces in the city, though in a localized manner, and those areas require a 'relational' and 'pluralist' understanding (Pieterse 2008: 106) in order to better grasp the relations between informal settlements and the formal vision of urban order,

[5]From a historical perspective, Freund analyzes the era of globalization and the consequent exclusion of Africa from "global cities" (Freund 2007: 171) in contrast with Afro-pessimism. He underlines the positive aspects of African cities and their creative and original responses to changes in the global economy, pointing out that some of the changes that are occurring in other world cities are also occurring in Africa, and the possibilities for transformation are considerable. They contain 'the seeds of new types of urban forms and solutions' (Freund 2007: 172), and are not merely the helpless victims of colonial and post-colonial attempts to implement modernist plans. Globalization is not, therefore, something that 'strikes' Africa from outside, rather it works within Africa in a way that suggests future models. Some strategies for pursuing development are clearly exclusive in this context, but the alternative solutions that people develop in order to carry on with their lives, which are increasingly 'urban' (according to forecasts), are the most unpredictable. In this sense the future is open, uncertain, and not without contradictions (Freund 2007: 172).

knowledge that is necessary in order to define directions and actions that are of concrete use in improving certain conditions.

One of the most interesting and distinct characteristics of this new and growing literature on urban Africa is its vast and valuable intersection of urban theory and planning practice (Myers 2011: 15). The theoretical attention paid to settlements that are marginalized and informal, invisible, necropolitan or ordinary in various African cities (Myers 2011: 14) is important but insufficient when constructing a new interpretive approach. Urban studies must also address *practice* as well as the subsequent attempt to define how the urbanization process can contribute to improving the quality of life of inhabitants (Roy 2007).

2.1.2 Urban Bias: Legacy and Continuity in the Conceptualization of Development and the Rural–Urban Relationship

If one criticism of African urban studies is that most of the analytic approaches adopted apply models of urban development based on the characteristics of major Western cities, another limit derives from the continued use of analytic categories and differentiations developed in specific historical and territorial conditions that may no longer be relevant to contemporary circumstances. One example is the tendency in the academic literature to divide the world economy into the simplified spatial categories of central and periphery, which is often linked to similar tendencies to draw a clear distinction between 'peripheral urbanization' (cities of the Third World and urban peripheries) and 'central urbanization' (cities of the First World and urban centres) (Murray/Myers 2006: 9). These categorizations negate a nuanced interpretation of the urbanization process, and ignore historical specificity, social complexity, and the differences between various cities and within the same city.[6]

Cities have been viewed from time to time as the product of economic change, motors of modernization and obstacles of development. These visions of the 'urban' and the related interpretations of the relationship between urban and rural spaces have strongly influenced the interpretive approaches to African cities and the rest of the world, and particularly the view of environment and natural resource management in urban and rural contexts.

The development debate has directed those interpretive approaches, and has developed mainly around changes in the relationship between agriculture and industry and the distribution of investments between various sectors. The goal of

[6]"Abandoning dichotomies (such as the rural-urban divide) allows us to fight the tendency, generally present in both academic and popular literature, to group all African cities under the same generic umbrella of examples of the Third World or peripheral urban planning" (Murray/Myers 2006: 9).

development policies has almost always been economic growth (Tacoli 1998b: 149–150).[7]

2.1.2.1 Urbanization as a Product of Economic Change

During the post-war period, cities came to be viewed by modernist theorists as the natural consequence of economic growth and investment in urban infrastructure and industry. The economic development model based on the 'traditional' (rural) agricultural sector was replaced by 'modern' (urban) industrial sectors in a growing economy. Until the mid-1960s, rural–urban migration was encouraged by the lack of manpower in urban areas and the possibility of keeping salaries low (Lewis 1954; Tacoli 1998b: 150). In Tanzania, the post-independence migration from rural regions to Dar es Salaam was, and still is, attributable to the economic and social opportunities offered by the city.

The expansion of modern industry, guided by key economic sectors, was linked to spatial concentration and inequality, resulting in irregular development and polarization, which were seen as the inevitable condition of countries in an initial state of economic growth. Such imbalances were considered solvable through the distribution of benefits in the surrounding urban area (Hirschman 1958, in Beall/Fox 2009: 21–24), but difficult to overcome once they had been established (Myrdal 1957 in Beall/Fox 2009: 21–24).

At the time, regional planning theorists and professionals thought that spatial disparities could be mitigated through targeted interventions. Policies were theorized that would encourage the growth of medium-sized cities in peripheral regions and help develop poles of urban growth (Friedmann/Alonso 1975), maintaining that urban centres could be used to guide regional development in nations with less-developed economies. At the beginning of the 1950s, the term "over-urbanization"[8] (Hoselitz 1957; Sovani 1964; Rakodi 1997) was linked to the emergence of concerns over the level of urbanization in Latin America, Africa, and Asia. Cities were increasingly characterized by the development of squats, shantytowns, and favelas, fuelling fears of the social and political impacts of urban growth.

---

[7]The Rome Club has been developing a variety of theories and manoeuvres for degrowth (e.g. transition towns) since the 70s (Illich 1994; Latouche 2005; Bateson 1976; Bologna 2004; Brown 1980).

[8]Over-urbanization refers to the relationship between the national level of urbanization and work force distribution across various sectors (e.g. agriculture, manufacturing, industry, and services). A country would be defined as over-urbanized (or under-urbanized) if the relationship between the percentage of the urbanized population and the percentage of the work force employed in industry was significantly different when compared with the same relationship in advanced economies. At the 1956 UN conference it was declared that in over-urbanized countries, urban and rural poverty exist one beside the other (Sovani 1964: 113, in Beall/Fox 2009: 20).

Such concerns were amplified by dependency theories that criticized the fact that development was inevitably related to inequality, unbalanced growth and the 'natural' evolution of the urban system over time. They maintained that development in the towns and cities of less economically advanced countries favoured Western economies rather than local areas. Cities were seen as commercial centres in an international system where primary goods moved from rural producers to international metropolises, via the markets in cities, regional centres, and capitals, creating a dependence pattern defined as 'parasitic urbanization'.

2.1.2.2 Urbanization as an Obstacle to Development

At the end of the 1970s, rural–urban migration was no longer seen as an economic incentive and rapid urbanization became a problem.[9] The negative view of the city as a parasitic and dystopic space that prevented development, as opposed to the motor of regional and national economic modernization, was solidified with the urban bias thesis (Lipton 1977; Bates 1981; Argawala 1983).

Adapting Lipton's theory to Africa, Bates (1981) emphasized not only the economic dimension of the problem, but also the superior political power of cities as compared with small farmers. He argued that Nyerere's government was perpetrating urban-biased policies directed at privileging urban dwellers with favourable urban wages and food subsidies relative to rural peasants' agricultural commodity prices.

After the support given to the Tanzanian state in the 1960s and 1970s, many Western donors, led by World Bank and the International Monetary Fund, criticized the role of the state as the central agent, the over-centralization of government and parastatals in Dar es Salaam, and the accountability of bureaucratic agencies. Charges of corruption and incompetence against governmental and parastatal agents were aligned with the theory of urban-bias (Lipton 1977).[10]

The 'green revolution', begun in the 1970s, sought to increase agricultural productivity in low- and medium-income countries and to introduce more productive crop varieties and improved agricultural practices. Those debates and policies led to strategies for managing rural–urban migration in order to prevent over-urbanization. Improved control of urbanization was expected to improve the living standards in rural areas and cities. If agriculture became more productive and if adequate social and recreational services were provided in rural areas, people would not move to the

[9]"It was clear that the creation of employment in the manufacturing sector was lower than expected and could not absorb the rapid growth of the urban population. Concerns regarding over-urbanization translated into policies that sought to limit workers' migration towards the city. At the same time, the first studies on the informal sector (Hartm 1973; ILO 1972; Weeks 1973) joined the debate already under way on the development potential of that sector (Portes et al. 1989; Moser 1978; Standing/Tokman 1991)" (Tacoli 1998: 150).

[10]The World Bank published its version of the urban bias theory and a critique of the bureaucratic underperformance of the African modernizing state.

city. However, Todaro (2000) reinforced the political implications of Lipton's urban bias thesis arguing that rural–urban migration was rational because it was based on expected rather than actual benefits, and that this was inevitable due to the imbalance between rural and urban economic opportunities in the majority of low- and medium-income countries.

These ideas were in line with the opinions of many of the leaders of low- and medium-income countries of the period, including Julius Nyerere in Tanzania, who sought to address the problem of poverty among farmers by reducing the differences between rural and urban welfare (Beall/Fox 2009: 21–24).

While international dependency theories achieved notable success during the 1970s, a liberalistic 'counterrevolution' began to appear which eventually dominated development studies during the 1980s and 1990s. The Structural Adjustment Programmes, although they were introduced to reduce the discrepancy between rural and urban incomes (and therefore to reduce rural–urban migration), did not allow equal access to international markets for all producers, which aggravated social disparities in the city and the countryside. In Tanzania, the social services and subsidies initiated under Nyerere were gradually eliminated, creating pressing cash needs for peasants (Bryceson 2002). Migration as a survival strategy, together with diversification of income, has therefore continued to be an essential element of livelihoods and accumulation strategies for those who live in the rural–urban interface (Tacoli 1998: 151).

The lack of consideration for cities that derived from the diffusion of the urban bias thesis beginning at the end of the 1970s is reflected in the *Poverty Reduction Strategy Papers* (PRSPs), which predominantly emphasize the relative importance of rural poverty and development, while giving less consideration to urban poverty (Tacoli 1998).

Given the demographic trend of rapid growth, the inevitability of urbanization, the undeniable importance of cities in economic development, the obvious increase in poverty and urban inequity, the continued influence of the urban bias thesis is difficult to understand for some authors. Jones/Corbridge (2010) have highlighted its limits, arguing that the urban bias thesis can be redefined without reducing it to a generalized model of city–countryside exploitation, which would be misleading if applied to individual urban or rural social classes. The urban bias thesis has promoted limited consideration for urban poverty and economic dynamism in many cities of developing countries, generating an imbalance between rural and urban areas in the political sphere, an imbalance that should be levelled (Jones/Corbridge 2010: 16). However, the concentration of economic and political power can offer advantages when constructing dynamic regional economies and livelihood strategies based on mobility.

Urban bias was also questioned because it neglects the organization, objectives, and expression of political institutions, which may have implications for the power and well-being of the rural sector. A conception of rural interest that is strictly focused on economic issues was also considered a limit of urban bias theory, arguing that the cross-cutting nature and complexity of rural cultural identities and interests may weaken the countryside more than the power of the city (Varshney 1993).

Moreover, the complex interactions between rural and urban economies and governments, and the arbitrary way in which urban and rural are distinguished were used to question the rural–urban comparisons of the urban bias theorists, and a 'large city bias' or "capital city bias" (Hardoy/Satterthwaite 1989) was proposed to label the mainstream government policies subsidizing consumer, business or land owners in capital or large cities.

2.1.3 The Rural–Urban Relationship and Politics of Development: Approaches to Environmental Planning and Management of the Peri-urban Interface

In 2008, the population of urban areas exceeded that of rural areas for the first time in history (UN 2008). The 2014 revision of the *World Urbanization Prospects* indicated that 3.9 billion people, or approximately 54 % of the world population (expected to increase to 66 % by 2050), live in cities, while rural population was close to 3.4 billion, and is expected to decline to 3.2 billion by 2050 (UN-DESA 2014). Urban population growth is concentrated in the less-developed countries where it is estimated that by 2017 a majority of people will be living in urban areas (UN-DESA 2014).[11]

It is expected that modalities of growth will vary considerably between continents and settlements, both in terms of typologies and dimensions. Africa, as well as Asia, is urbanizing faster than other regions and is projected to become 56 % urban by 2050. In the countries of the Global South, a growing portion of the population lives in the areas surrounding large cities. In such areas, which have hybrid rural–urban characteristics, living conditions depend on local natural resources, such as arable land, water, fuel, and living space (Tacoli 1998).

Spatial policies (regional and local) are generally used as instruments for obtaining a better balance between city and countryside and for reducing migratory pressures on urban centres. The failure of such policies is often attributed to a lack of recognition of the complexity of rural–urban interactions, which involve spatial and sectoral dimensions (Tacoli 1998: 3).

Research on planning and management of areas with hybrid rural–urban characteristics in 'developing countries' have been mostly oriented towards poverty reduction and have sought to directly meet the needs of the most vulnerable, or to render more sustainable the livelihood strategies based on the use of land and natural resources (Mattingly 2009: 38; Simon et al. 2004). Power dynamics and

---

[11]The global urban population is expected to grow approximately 1.84 % per year between 2015 and 2020, 1.63 % per year between 2020 and 2025 and 1.44 % per year between 2025 and 2030.

relationships of cause and effect that reflect non-homogeneous socio-economic and environmental realities remain little explored. Studies that have concentrated on the 'negative' aspects and problems of the development dynamics in cities (poverty, precarious health and environmental conditions, etc.) have paid little attention to the structural causes of peri-urban settlement formation and their unsustainability and vulnerability. Such causes include the expulsion of the population from central areas of the city in search of more liveable areas and more accessible resources (land to cultivate, water that can be drawn directly from natural sources, etc.) or competition for the right to use the areas most favourable for tourism development.

2.1.3.1 Neither Urban nor Rural: Definitions

A variety of perspectives can be found in the literature on areas that can be defined as neither rural nor urban. From a morphological-landscape point of view, attention has been directed to low density and apparently random, scattered or fragmented and discontinuous forms of land use that can neither be classified as urban fabric nor as true rural areas. From this perspective, such areas are described as a 'type' or operational dynamic of land use, such as zones of separation between the city and the country (the urban fringe theory), or the dynamic and rapid transformation of rural areas into urban ones (the urban sprawl approach). In both cases, areas with hybrid rural–urban characteristics are viewed as 'anomalies'[12] of the urbanization process and as the evolution of rural areas. Some authors, however, consider rural–urban transition zones to be a specific environment. In some cases, it is recognized that such areas have their own 'identity' as peri-urban (or semi-urban) as a "distinct ecological and socio-economic system under an uncertain institutional regime" (Allen 2006: 32). In other cases, following the suburbanism approach defined by Ekers et al. (2012), they are considered an expression of the global suburbanization process, "the combination of non-central population and economic growth with urban spatial expansion".

The approach adopted in the present research considers peri-urban areas as having their own specificity; however, in this book this distinction should be understood not as a defining approach that lays out conceptual boundaries, but more as an explorative (inquisitive) approach. Peri-urban areas are considered the location of a peri-urban way of life, which suggests a reframing of the question of the urban in sub-Saharan Africa.

In order to overcome a conceptualization that is centred mainly on physical aspects (distance from urban areas, density or infrastructure), the present work refers to a gradient between urban and rural poles. That gradient can only be understood by examining the dynamics of interaction in the rural–urban interface (Simon et al. 2001), by paying attention to socio-economic heterogeneity and the complex

[12]Anomaly is understood in this case not as an extraordinary phenomenon, but as an imperfection and a deformation of urban or rural areas.

institutional balance (Tacoli 1998, 2003; Mattingly 1999; Allen 2003), by examining the mosaic of 'natural' ecosystems, 'agricultural ecosystems', and 'urban ecosystems' involved in the flow of materials and energy required in urban and rural areas (Morello 1995) and by analyzing the system of governance (Ekers et al. 2012).

From this perspective, the rural–urban interface represents the context in which many changes, flows, and rural–urban interactions materialize, generating problems and opportunities not only for the communities that live in the interface but also in terms of the sustainable development of adjacent rural and urban systems. In this last sense, although the framing of the debate has changed considerably since Lewis' time, the correlation between rural-poor and urban-rich, as well as approaches that reconfirm this dichotomy in the theoretical sphere, in conceptualization, policies, and practices (institutional), are still present. Moreover, suburban theorists consider the suburbs "as an area's constellation of public and private processes, actors, and institutions that determine and shape the planning, design, politics, and economics of suburban spaces and everyday behaviour" (Ekers et al. 2012; Mabin et al. 2013; Buire 2014; Todes 2014).

2.1.3.2 Planning the Peri-urban Interface: Urban Fringe or Transition Zone Between Rural and Urban?

These analytical approaches inform policies and plans. From a planning and management perspective (for development), two distinct lines of thought exist (Budds/Minaya 1999: 21). Both involve an approach that considers the specific characteristics of rural–urban areas, and define them as the peri-urban interface (a debate that has been advanced considerably through the *Strategic Environmental Planning and Management for the Peri-urban Interface* project, undertaken by the University College London's Development Planning Unit).

One of these lines of thought prevails among international agencies, such as USAID, UNICEF, the World Bank, and CARE, as well as in the rural planning and management manuals for the area (e.g. Dala-Clayton et al. 2002; Mattingly 2009: 38). Their interpretation assimilates peri-urban areas to 'slums' (drawing from urban fringe theory), meaning settlements characterized by low-quality housing, insufficient levels of infrastructure and services, and no recognition of the legal right to occupy land (Beall/Fox 2009: 27). This simplification, in addition to concealing the socio-economic diversity and variety of settlements found in peri-urban areas, fails to account for the resources that such areas provide to their own inhabitants and to the entire city.

A second interpretation dominates the activities of other agencies and programmes that are more attentive to questions of environmental management and conservation of agricultural land (e.g. the Natural Resources Systems Programme, the IDRC through the Cities Feeding People Programme, and the FAO) and considers the peri-urban interface a transition zone between rural and urban areas (Budds/Minaya 1999: 21).

There is a significant difference between the priorities and intervention strategies of these two approaches. The agencies that limit themselves to considering peri-urban areas as fringe areas tend to support interventions with an urban perspective, generally concentrating on lack of infrastructure, provision of potable water and sanitary systems, education, and community health. Agencies that have a more environmental focus recognize that peri-urban areas also contain rural zones near the city limits. As such, they concentrate on the management of natural resources, including urban agriculture and the effects of pollution, and promote the development of more sustainable practices rather than the development of urban infrastructure.

2.1.3.3 A Common Strategic Element: Community Participation

Most development agencies demonstrate a strong commitment to community participation as a management strategy for peri-urban areas. Over the course of the debate on development and environment, participation has emerged as a method of expressing the needs of the most vulnerable. Having a voice in the political sphere represents a central element for improving one's quality of life. A report by the British Department for International Development[13] on the challenges of urban development in poor contexts (DFID 2001), emphasized the diversity of people's needs and the importance of their participation in identifying which services they need, where such services should be located, and how the potential benefits should be distributed. The report also highlighted the importance of having the capacity/possibility to participate in decision-making processes, and to play an active role at every level of the development process, in a manner that allowed people to identify and monitor the improvements that they felt were needed. Specifically with respect to environmental planning and management in peri-urban areas (see Sect. 3.1.1), the criteria for participation refer to the collective use of individual powers. Addressed collectively, changes can be managed in a more practical way, and the impacts of environmental transformations affect everyone within a given area, therefore each person will benefit from improvements (Allen et al. 2003). Different actors have varying capacities and strengths, some contributing with practical knowledge, others with institutional knowledge, and others still with financial means or labour. A successful process must be defined on the basis of all these potential contributions in order to achieve sustainable development (Allen et al. 2003).

---

[13]Here we refer to the assertions made in the agency's Strategy Paper, *Meeting the Challenge of Poverty in Urban Areas. Strategies for Achieving the International Development Targets (2001: 24)*, as an example.

2.1.3.4 Peri-urban Livelihood and the Inevitable Transition to Urban Status

In the sphere of the livelihood approach, developed mostly in the early 2000s and focused on environmental management and local community participation, the centrality of the livelihoods in the peri-urban interface (Simon et al. 2004) led to renewed interest in urban expansion as a phenomenon that could influence the availability and accessibility of resources (means of subsistence) (McGregor et al. 2006). In most cases, this interest has translated into a focus on urban and peri-urban agriculture, in response to cities' increasing food demands, and practiced in areas where non-agricultural use of land is also an option, unlike in rural agriculture (Maxwell 1999).

Although this focus addresses the issues of land and food security and may be useful for environmental management, it does not consider the complexity of people's livelihood strategies (Mattingly 2009: 38). An agriculture-centred discourse on peri-urban areas may have important implications and political consequences. Although agriculture could be practiced for long periods of time by farmers who are already integrated into the urban economy, Mattingly (2009) argues that peri-urban agriculture in developing countries is an inevitably temporary activity in nearly all cases. It is located in areas that will eventually assume urban uses, since those who practice peri-urban agriculture are in transition from a rural economy to an urban one. Moreover, peri-urban agriculture in developing countries could be practiced by the poor as a livelihood activity, but also by individuals with higher income levels as a supplementary activity. In any case, the changes in land use (and in the environment more generally) can have considerable impacts on the livelihoods of people who live in the areas defined as peri-urban. As such, policies and interventions for better environmental management, local food security, and support for small farmers (especially related to land) must consider the effects that they will have on people's lives and livelihoods and support the 'transition' of people living in peri-urban areas from a rural life to an urban one (Mattingly 2009: 50).

At this point a problem of planning competence arises, around which institutions should develop plans sensitive to the social and economic characteristics of peri-urban areas.

Some authors maintain that policies and projects for natural resources may not reflect local priorities near urban areas, since income generating activities (Brook/Dàvila 2000) that are not based on natural resources become more important, and may be undervalued by political decision-makers who are more oriented to solutions focused on natural resources. A rural administration concentrates mainly on rural environmental issues that affect the majority of its residents; however, with the rise of urban development, they need to be more oriented towards the urban. Local rural administrations can do more to support agriculture in the areas adjacent to the city by providing technical consulting (to improve agriculture and soil fertility, access to credit and security of property) and doing more to protect the land rights and the interests of farmers, including those that conflict with more powerful and aggressive urban administrations (Mattingly 2009: 50).

Urban administrations are equally responsible for the impacts of their actions. They are responsible for the new demand for land and waste management, often extending their territorial planning to rural areas and beyond municipal boundaries. Their rapid acquisition of large areas may destroy rural livelihoods. According to Mattingly (2009), they should plan city expansion in a way that allows vulnerable farmers more time to undertake rural–urban transition, and should act directly or indirectly to provide better compensation for the land rights of which such people are deprived. According to Mattingly, in both rural and urban institutions, new goals, knowledge, competencies, and experts are necessary for improved transition to urban economies, improved food security for a larger number of people, and possibly also for the management of rural and urban migration in general. Providing support for changing livelihoods through appropriate policies and actions related to land could help people to better face the difficulties related to urbanization processes, to maintain or increase food and income sources and develop competencies that are of value in urban economies.

The extent to which the transition to a totally urban environment is the inevitable destiny of all peri-urban areas is a point of discussion in the present study. Does the shift from rural to urban occur in any case, and is people's end goal to settle in peri-urban areas? Or do such areas represent a hybrid space that continues to straddle economies, forms, spaces, flows, and relations that are both urban and rural, while at the same time neither urban nor rural? How do people address the anthropic and natural environmental changes that are occurring in such areas: with an attitude of resistance, or flexibility/adaptation? Why?

The assumption outlined above often translates into policies that support transition and that provide security of property and agricultural support, and this begs several questions.

If access to land in these areas represents, in the majority of cases, an important component of livelihoods, is it equally true that security of access to land corresponds to possession of a land title? What are the modalities of access to land and resources? What relationship is there between so-called 'informal' and 'formal' modalities?

2.1.3.5 Peri-urban as the 'Black Periphery' and the Emblem of Suburbanization

Widening the analytical perspective of African peripheries (and peri-urban areas) in his book *City life from Jakarta to Dakar. Movements at the Crossroads*, Abdoumaliq Simone investigates the reasons why what is defined as 'periphery' represents an extremely important place for urban life (cityness), and encourages us to take a step back in our reflections on urban life (Simone 2010: 14). Independently from the relationship between conceptualizations of peri-urban areas and the periphery, the author describes peri-urban areas as one of the five manifestations of periphery, thus providing interpretative elements that allow for a broadening of the conceptualizations that have been outlined above. Defining peri-urban areas as places of transition and connection that function in a variety of ways, Simone identifies four main characteristics of those areas: they are places where migrants first settle before moving to the city centre; they are territories of agricultural production contributing to subsistence in peri-urban areas themselves and providing products to urban markets; they are the areas to which nearly all environmental costs are exported (i.e. waste, polluting industries, and congested transportation hubs); they are boundary areas in which temporary, residual or new links with the city and with rural areas are maintained and intersect (Simone 2010).

Particularly in Africa, the impact of city expansion and the demand for resources in surrounding areas competes directly with the use of rural land. In such areas, urban residents maintain ties to rural existence as a socio-cultural tendency. Access to employment, housing and sociality in urban areas are often subordinate to the way urban residents relate to specific rural resources, such as land, rural livelihoods, and local policies. In other words, what urban residents are able to do in the city (for example how they find a house, or how they work in cooperation with others in order to build their livelihoods) may sometimes be a function of how they position themselves with respect to historic rural connections (Simone 2010).

These reflections highlight how urban and rural areas are connected by complex moral economies that require both a clear distinction and interconnections between the two domains.

The main challenge is to determine how rural and urban residents, households, and communities can support each other without the occurrence of abuses of power. This kind of reciprocal relationship requires, according to Simone, "that people imagine themselves living in a translocal topography that incorporates the rural and the urban not as clearly defined and opposed domains, but fractured ones, with different connotations, expectations, practices and strategic orientations" (Simone 2010: 53–54). From this perspective, peri-urban areas are also a mix of temporality. This means that it is difficult to understand what kind of development is occurring or is about to be undertaken, what is increasing and what is declining (Simone 2010: 54). In most cities, architecture, infrastructure, and land development have been used as instruments to induce new institutional and social urban relationships in relation to how decisions are made, what is considered possible or useful in the city, and how financial responsibilities must be defined and risks assessed (Simone 2010).

Shifting to a political ecology perspective, a recent set of studies on suburbanism that focus on suburban governance[14] argue that 'a suburban perspective' is very important to this shift in perspective on the modalities of governance in the peripheral city (Ekers et al. 2012).

---

[14]They propose three distinct but complementary modalities of suburban governance: the state, capital accumulation and private authoritarianism to critically discuss the governance of suburbanization and diverse ways of suburban life.

Conceiving suburban governance as "the constellation of public and private processes, actors, and institutions that determine and shape the planning, design, politics, and economics of suburban spaces and everyday behaviour" (Ekers et al. 2012: 406), they include peri-urban areas among the different suburban forms of urban decentralization. On the one hand, they highlight how "powerful processes of uneven development, capital accumulation, migration, and agricultural transformations have resulted in varied forms of peri-urban development that touch all urban-regional spaces" (Ekers et al. 2012: 406). On the other hand, they discuss power relations, inequality, and marginalization, which profoundly affect the trajectories of suburban growth and decline, influencing social and ecological histories of suburban transformations and forms of everyday life.

2.2 Agency and Environmental Management Practices

The relevance of social relations in the informal sphere (as the primary means of city development), and of how "micro policies define local dynamics and the conditions of the unprivileged" become clear in the debate on African cities (Lourenço-Lindell 2002: 21–22). Attention should be paid to the social bases of informal livelihoods, as they are the forms of social organization and interaction that support and limit the life strategies of people living in peri-urban areas.

Although it does not enter into the particulars of the nature and composition of such social networks, the analysis of practices for management and access to resources, and for adaptation to environmental change in peri-urban areas, has indicated the extent to which such practices and strategies are based on those relationships (and determine the capacity to act). The dependence on natural resources and the structure of such social networks therefore assumes a central role in terms of the capacity to adapt to environmental change.

Life, and in some cases survival in cities (and especially in informal settlements such as peri-urban areas) is based on 'face-to-face' interactions,[15] while one can rarely trust institutions that have few resources and an impartial approach to resolving conflicts and to relationships in urban society (Lourenço-Lindell 2002). Networks of personal relationships are the instrument through which people can access living space, land to cultivate, credit, information on prices, and assistance in difficult circumstances, and are therefore the basis of their livelihoods.

The importance of social networks and relationships has likely taken on more importance as a result of the 'urban crisis', specifically the criticalities evinced by the debate on Global Climate Change in African cities, extreme weather events (tsunamis, flooding, etc.) and the stresses to which cities are subject. In this difficult

[15]Although there has been a considerable increase in the diffusion of and access to communications technology, due to poor infrastructure many African cities are still fully or partially excluded from the 'network society' at the centre of Castells' 'information capitalism' (1998: 92–95, cited in Lourenço-Lindell 2002: 22). Nevertheless, they do constitute another type of 'network society'.

environment, where there is a constant lack of resources and uncertainty, other kinds of rights must be activated to create opportunities to address this crisis and reduce vulnerability. Through the creation of links with others, people generate reciprocal expectations, developing needs and creating new rights and rules which govern relationships and orient behaviour.

2.2.1 Vulnerability Is not Inherent

A variety of debates have developed around the question of 'assistance' relationships. In particular, two broad debates have arisen regarding the concepts of *vulnerability* and *livelihoods.*[16]

The first of these was born of criticisms of conventional approaches to poverty as regards their excessive focus on variables related to income, and their undifferentiated and passive view of the poor. At first, the term *vulnerability* was proposed as a conceptual instrument that would allow for a broader consideration of the dimensions of hardship, including isolation (Chambers 1989; Rakodi 1995; Watts/Bohle 1993, in Lourenço-Lindell 2002: 22). Unlike poverty, vulnerability does not mean lack or want, but defencelessness, insecurity, and exposure to shocks and stresses (Chambers 1989: 1). As such, material poverty cannot necessarily be seen as the only source of vulnerability, nor can all poor people necessarily be seen as equally vulnerable. From this perspective, connectivity (the total of social relations) takes on a central role in the definition of people's 'well-being' (quality of life).

The Livelihood Framework was also developed to understand people's many activities and the modes in which they live. In this framework, livelihood components are people's capabilities, and the tangible or intangible means that they have at their disposal.

The present research pays particular attention to the 'intangible' dimension that consists of needs (for concrete means or support) and the exercise of informal rights and rules that are created to support life in the urban and peri-urban sphere. In informal contexts considered lacking in infrastructure and services, one wonders how that dimension contributes to supporting or integrating the role of formal (institutional) infrastructure and services, and how they, in turn, define the relationships between people who live in peri-urban areas, environmental management, and access to resources. Individuals' and communities' capacities to address environmental changes, and to develop livelihoods that allow them to mitigate or avoid the negative impacts thereof, depend on precisely these relationships, and on the informal and 'intangible' production of infrastructure and services.

The tangible and intangible dimensions of the Livelihood Framework therefore become the elements that determine vulnerability to environmental change,

[16]As old and new variants of the debates on 'informal security systems' and the tradition of 'social networks' (Lourenço-Lindell 2002: 22).

specifically autonomous adaptive capacity in peri-urban areas. The objective of this research is to discuss, on the basis of the results of the data analysis, a spatial homogenization in which urban means rich and equipped with infrastructure (if slums are excluded), while peri-urban is a synonym for periphery, marginal, poor, and 'lacking'. In the hopes of shedding light on that homogenization, the analysis of several peri-urban areas of Dar es Salaam has been conducted in order to identify the elements that express their environmental and social complexity. This research assumes that vulnerability is not intrinsic (and that not all poor people are equal), rather it is attributable to the conditions of a given context, which depend considerably on the physical environment and on social networks, in addition to the intrinsic characteristics of people or groups of people (see Chap. 4).

2.2.2 *The Role of Agency in Informal Settlements*

Nevertheless, this study is not a mere application of the Livelihood Framework, and in fact does not provide exhaustive documentation of the components of the livelihoods of households living in peri-urban areas. It is focused, rather, on a few elements of informal (local) livelihoods, and on the intangible practices and relationships that develop around them (modalities of accessing and managing natural resources, actions and means of autonomous adaptation to environmental change).

The debate on the capacity to work and to develop livelihoods in the informal sector, which consists of relationships that are broadly developed in a variety of disciplines, oscillates in some cases between a celebratory approach and a victimization of the individuals in those areas, which often fails to consider the role of people as agents of spatial transformation and management and as political actors (Lourenço-Lindell 2010).

On the basis of such limitations, a variety of approaches have been developed that have given a central role to the agency of people in terms of the economy and in informal settlements. Some of the literature has addressed the living conditions in the urban sphere and strategies of income production, emphasizing the diverse modalities with which people do their best to face difficulties such as the contraction of the formal employment market (Rakodi/Lloyd-Jones 2002). Other authors have stressed the various typologies of social networks through which people support their informal activities and income production in crisis situations (Lourenço-Lindell 2002).

Recent studies have pointed to the capabilities and opportunities generated by temporary and new relationships and extended forms of collaboration between people with different cultural and social backgrounds in urban African areas (Simone 2004: 10–13). From this perspective, water networks and land access modalities can be seen as 'platforms' in which people collaborate in a 'silent' but effective manner (Simone 2004). Moreover, daily social practices and informal networks are, in some cases, intentionally disguised and concealed so that the State will be unable to distinguish and govern them. In this way, people are able to resist government decisions and collaborate through 'tacit power' (Simone 2004). In a broader sense, other researchers are paying increasing attention to local (informal) practices, not as marginal manifestations of chaos and decline nor as a deviation from normative Western ideals, but as the basis of a social system in which a different kind of urbanity (city) is possible (Simone 2004; Pieterse 2008). Daily informal practices are viewed from a political perspective as 'insurgent citizenship' (Pieterse 2008), since they are the actions through which people live and interact with the urban space in order to meet their needs, as is widely evident in cities (Lourenço-Lindell 2010).

2.2.3 Human Agency and Power

Although some authors maintain that the theoretical framework of human agency, like the livelihood framework, can help to clarify conflicts and contradictions in peri-urban areas (Mbiba/Huchzermeyer 2002: 122), critical aspects and limits of human agency theory have also been highlighted with respect to the acquisition of power, the means to act or not act, and the capacity to obtain results and influence events. Those authors maintain that although various studies of livelihoods and changes in peri-urban areas involve an understanding of power through their focus on capacity building and governance (Rakodi 1998; Nittingham/Liverpool 1999, in Mbiba/Huchzermeyer 2002: 122), there has been an ambiguous use of the concept of power when exploring the contradictions and conflicts in peri-urban areas. They suggest that analyses should explore the impacts of the responses adopted by institutions and by various socio-economic interests, and the links with changing conditions in peri-urban areas. Beyond the generally convergent variety of approaches used in the study of change in peri-urban areas, one must also bear in mind the way in which they are mediated by contextual variables such as place, time, history, scale, and diversity of actors. Thus arises the need for research on peri-urban areas that recognizes these contextual factors, as well as the relative issues of language, culture, governance and power (Mbiba/Huchzermeyer 2002: 127).

Other authors criticize the agency approach because the combination of the population's limited power, the weakness of formal institutions, and the decentralization of participatory approaches has favoured influence peddling, rather than promoting democratic forms of governance and relationships of transparency and monitoring between the state and society (Meagher 2010, in Beall et al. 2010: 197). Meagher, hoping for less descriptive and more analytical approaches, maintains that the density of social connections is less important than what networks actually connect to for city inhabitants. For example, an abundance of solidarity networks that connect poor people to each other in a horizontal manner can be associated with a lack of resources and political influence. Networks that require or generate links with more influential segments of society are often more effective. Referring to the work of Lourenço-Lindell (2002) and Simone (2001), Meagher suggests that a web of horizontal connections and associations can 'exacerbate rather than reduce vulnerability'. The political voice of the poor runs the risk of being limited by the impossibility of connecting with formal government (institutions), by their lack of social and economic power, and by the consequent failure to represent their own interests within the governance structure. As regards defining policies for supporting the construction of livelihood networks and community-based solidarity associations, efforts should be made to draw associations closer to institutions and government actors. This means that promoting what is defined as "participatory development", which is not just social relationships and self-help, necessitates greater attention to power (Beall et al. 2010: 198).

As Roy (2005: 148) has asserted, the celebration of self-help in the wake of the work of authors such as De Soto (2000) or Hall/Pfeiffer (2000), runs the risk of obscuring the role of the state to the point of rendering it unnecessary, and actually reaching the point in some cases of legitimizing privatization through neocommunitarianism models. On the other hand, in the interaction between informal networks and formal systems of government, imbalanced power relations, and exclusive systems of land management can be generated that should be carefully evaluated.

2.2.4 Building the City: 'People as Infrastructure'

Some African cities (e.g. Lagos) are used as models for the future due to their capacity to function despite an apparent lack of coordination or planning. However, some think this approach risks condemning many of the populations of African cities to perennial disadvantage and poverty (Murray/Myers 2006: 237). Upon closer examination, African cities are beginning to develop their own vision of African urbanization (Odendaal 2012; Ngau 2013) through which an authentic dialogue can be opened that reflects the experiences of other rapidly growing cities in the Global South, bringing the African city to the fore in terms of decision-making and political debate.[17]

Abdoumaliq Simone's extensive study of African cities, which combines urban planning and the postcolonial literature, has viewed African cities as examples of daily resistance against the inadequate responses of urban planners and development, rather than as failed cities.

In contrast with the vision of African cities as having fallen into ruin, Simone (2004b: 407) argues that these ruins constitute a highly urbanized social

[17]"The cities of the Global South have begun to take on an important role in urban planning theory, to the point that such cities no longer represent an anomalous category, but a fundamental dimension of the global experience of urbanization. A focus on cities such as Lagos has the potential not only to illuminate a peculiarity of the African experience, but also to answer broader questions on the nature of modernity, urban governance, and the interaction between flows of global capital and material conditions that actually exist in the Global South" (Gandy 2006: 250).

infrastructure capable of facilitating the intersection of social relationships and expanding spaces of economic and cultural operation. With the aim of revealing the potential of African cities and the ways in which their unique characteristics can be activated for the development of urban planning policies, Simone develops the concept of 'people as infrastructure', which emphasizes economic collaboration between residents who are apparently marginalized and impoverished by urban life. Infrastructure is commonly conceived in physical terms, as road networks, pipelines, and cables, which render the city productive by reproducing it and distributing residents, zones, and resources in specific formations in which the energy of individuals can be more effectively exploited and recorded. Simone hopes to extend the notion of infrastructure to people's activities in the city, characterizing the African city as a set of infinitely flexible, dynamic, and temporary intersections of residents (complex combinations of objects, spaces, people, and practices) who act without clearly defined notions of how the city should be lived or used. These connections become infrastructure, a platform for social transitions and livelihoods that allows for and reproduces city life. "This process of conjunction, which is capable of generating social compositions across a range of singular capacities and needs (both enacted and virtual) and which attempts to derive maximal outcomes from a minimal set of elements", is what Simone defines as "people as infrastructure"[18] (Simone 2004b: 410–411). This concept is crucial to the present research, which recognizes this type of infrastructure in peri-urban environmental and resource management and adaptation practices, and argues that they are a key point in analyzing African cities and informal urban planning and decision-making in sub-Saharan cities. Referring to the contributions made by Lefebvre and De Certeau, Simone argues that African cities survive mainly through a combination of heterogeneous activities exercised and elaborated through pathways that are configured in a flexible way. These flexible configurations are not pursued in opposition to urban priorities or non-African values, but as specific approaches to achieving the stability and regularity that non-Africans in the city have historically tried to achieve (Simone 2004b: 408).

His ethnographic work on the city of Johannesburg reveals that the growing distance between how Africans actually live and the normative trajectory of urbanization and public life constitute a new field of economic action (Simone 2004b). With limited institutional support and financial capital, the majority of urban African residents contribute, even if only a little, to collaborative processes and barely participate in the mediatory structure that prevents or determines how

[18]This notion seeks to extend what Lefebvre (1974) intended by the social space of practices, modalities of organization on a variety of scales, and connections that link expressions, attractions and repulsions, likes and dislikes, and changes and fusions that effect urban residents and their social interactions. The modes of doing and representing things become an increasing 'familiarity' with the other. They participate in a changeable series of reciprocal exchanges, in such a way that positions and identities are not fixed, nor in some cases can they even be determined. These 'urbanized' relationships reflect neither the dominance of a linguistic history or structure, nor a chaotic primordial mix (Simone 2004b: 411).

individuals interact with each other. This apparently minimalist proposal (bare life) nevertheless allows countless possibilities of combinations and exchanges that exclude all definitive judgements of effectiveness or impossibility. Throwing into the mix their profound particularisms of identity, origin, destination, and livelihood, urban residents generate a sense of unjustifiable movement that could remain geographically circumscribed or cover great distances (Simone 2004b).[19]

Simone (2010: 264–333) contrasts his position with the emphasis on control and sovereignty, and with attempts to revitalize the colonial present through new forms of militarization, which are cardinal characteristics of the present era that prevent one from paying attention to the modalities with which a great number of urban residents do more than "submit to the sentence of a simple life".

Simone's objective is to "provide a theoretical basis from which to promote a sense of 'multiplicity' in urban African development", which means "the ability to negotiate between knowledge of locally produced urban development and that which is produced externally, and to increase the influence of African experience and contributions in the consolidation of knowledge of urban planning in general" (Simone 2010: 241).[20]

2.2.4.1 The City of Black Urbanism

The need to recognize that 'non-cities' have led to their reclamation through the development of Black Urbanism, as a strategy that does not seek to identify a particular type of 'urban' but draws attention to several dimensions of urban life that are too often ignored (Simone 2010).

Black Urbanism is therefore a pretext for explaining how 'platforms' of engagement and collaboration can be constructed, beginning with the 'intricate' experience upon which the majority of urban areas of the world are based in various ways. Such experiences include the ways in which people live and share space and resources with their neighbours, construct and participate in networks of

[19]In this sense, Simone moves closer to those who criticize classical policies of separation between *bios* and *zoe* (e.g. Agamben). Illich, in a conversation with Rahnema said "the possibility of a city being an environment that fosters a common search for good has disappeared. [...] The drive for progress has extinguished the possibility of shared foundations for development. Irrespective of economic level, the good can arrive only from the type of complementarity that Plato, and not Aristotle, had in mind. To dedicate oneself to the other generates a unique space that allows for all that you ask: a mini-space in which we can agree on the search for the good" (extract from *Lo Straniero*, Year VIII—n. 45, March 2004).

[20]Simone (2010) has continued his work with the book *City life from Jakarta to Dakar. Movements at the Crossroads*, which connects African cities with Asian ones (Simone 2010: 14). He opens the debate by connecting the processes of African cities with those of other areas and develops the idea of "Black Urbanism" (Simone 2010: 268), drawing together a variety of situations and strategies that have worked in the long history of African people's movements in the 'urban world'. His use of the concept of 'blackness' is based on the hope that in the end it will be freed of its racial baggage.

relationships, evaluate their possibilities and keep various options open for the future, maintain the information rich environment that is necessary in order to continuously adapt their way of life, accumulate and manage resources, and overcome or manage specific difficulties.

Within this framework, 'people as infrastructure' can allow the cities of the South to have a "leading role in the creation of new synergies, transversal investments, commons, distribution networks, and multilateral alliances within which policies and key agreements are defined" (Simone 2010: 15–16), and offers a powerful means of re-evaluating humanity and citizenship without referring to the economic state of the African city (Simone 2010: 124–125). This leads Simone to partially redefine the "right to the city" as "the right to be chaotic (disorganized) and inconsistent, or to seem disorderly", conceived not as the right to be left alone, but to engage, to be an object in motion, and an actor of urban transformation (Simone 2010: 331).

The concept of self-organization in relation to that of 'blackness' becomes a device for connecting the various experiences of people of colour in the world (Simone 2010: 296–297). Some authors criticize this approach for using a "black geography that is too vague" (Myers 2011: 13), which is a useful contribution in terms of broadening the horizons of urban studies and connecting African and Asian cities, but is less relevant in terms of understanding more tangible elements related to what cities are able to do and why.

2.2.4.2 How to Reinterpret the Sub-Saharan City Through the Concept of 'Adaptive Capacity'

Considerable debates have developed on the main challenges that African cities must face (Myers 2011), for example regarding how to overcome the colonial inheritance of poverty, underdevelopment and socio-spatial inequity, and how to approach the management of informal sectors and settlements, governance and justice, creation of a non-violent environment, and globalization.

Despite the fact that many of the above-mentioned post-structuralist urban theories in African studies offer an innovative way to read city and urbanization processes, their alternative vision may still appear vague and difficult to interpret to many readers.

A more tangible way to analyze the complexities of what is going on in African cities is still needed (Myers 2011: 115). In particular, attention should be paid to the 'systemic drivers of urban development' including decision-making in urban politics, the infrastructure, technology, and landscapes of a city's spatial structure, and how the resulting inequalities are addressed (Pieterse 2008).

The question of how theoretical concepts of specific African cities can be combined with practical and experiential issues of urban life to alleviate social vulnerability to environmental change remains unanswered, and merits further consideration from economic, social, and environmental perspectives.

The question of environment, people's relationships with natural resources, and how those relationships influence and are influenced by urbanization processes has often been a minor aspect in the aforementioned debates. Within the urban bias thesis it was sometimes mythicized to the point of negating the legitimacy of an urban African environment, which was seen as the destroyer of the primitive man-nature utopia. In other cases it has followed in the wake of modernism and structuralism, developing around the sanitation and infrastructurization of the urban environment. In both cases, the dichotomy between rural and urban persists, as well as the impossibility of attributing hybrid rural–urban spaces, which occupy the interface between city and countryside, to their own non-transitory dimension.

The debate over and programmes for sustainable cities, initiated at the Rio Conference in 1992, shed light on the relationship between urban development and natural environments, and are relevant at the political level. However, in practice a dichotomous and sectorial Western approach continues to dominate urban planning. The debate on climate change that has developed in recent years has brought the environmental question to the fore, challenging both interpretative models and planning strategies and instruments at the global level, particularly in the areas that are more vulnerable to environmental change. The natural disasters and environmental stresses that we are now witnessing (the result of the combined effects of global environmental change and local anthropic activities, including urban development) necessitate new approaches to sustainable development (Simon 2007).

Cities, which are the causes and 'victims' of the environmental changes currently underway, are called on to respond through mitigation that acts on the causes of those changes and through adaptation to climate change that develops new spatial planning strategies (see Chaps. 4 and 5).

2.2.4.3 Towards Adaptive Strategies for the Sub-Saharan City

The goal of adaptation is to reduce the negative impacts of environmental change. Improving adaptive capacity in the peri-urban areas of sub-Saharan African cities is a priority for researchers because such areas are particularly vulnerable (Satterthwaite 2007; Tacoli 1998), and because attempts at environmental planning and management in those areas have proven to be ineffective in many cases (Friedman 2005). Although the approaches that have recently been theorized and applied are quite diverse (the resilience approach, the vulnerability approach, integrated approaches, etc.), they all seem to share certain assumptions that could lead to the implementation of inadequate strategies. In other words, a perspective that is heavily oriented by and constructed on the basis of Western epistemological and urban-centric models (cultural bias), which Robinson (2006) would call 'asymmetrical ignorance', is used in the interpretation of urban development processes and therefore in the evaluation of vulnerability and adaptive capacity. It is often assumed that all peri-urban areas are homogeneously poor or are transition zones destined to complete their process of urbanization. If those assumptions are not true of the context in question when identifying the local level factors that are determinative for adaptive capacity and defining the interaction between those factors, attempts to improve capacity may actually exacerbate the impacts of climate change and impede adaptive capacity. For example, considering agricultural activities as marginal or inevitably temporary can lead to the selection of adaptive strategies, such as accelerating the urbanization process, that may in fact render land access more difficult and may have negative impacts on people's livelihoods while also limiting their options for adaptation (see Chap. 4). In that case, rather than compensating people for the damages caused by environmental change (attributable mainly to the activities of Western countries) adaptation plans and measures would run the risk of causing further injustices generated by a lack of awareness of the dynamics in peri-urban and other areas that contribute very little to the emissions of greenhouse gases.

In contexts like sub-Saharan Africa, where the growth of cities is extremely rapid, settlement modalities are predominantly informal, and public and private funds are limited, it is essential that efforts to reduce vulnerability to environmental change be developed on the basis of people's autonomous activities. Beginning with local practices, and reconstructing what Simone defines the 'social infrastructure' upon which the strategies and actions that people undertake to address and adapt to environmental change depend, can shed light on the gap between the way people in African cities actually live and the normative trajectory of urbanization (Simone 2004: 407).

Providing the elements that will help to bridge this gap is the goal of contemporary research into the factors and processes that determine adaptive capacity in specific contexts, such as the peri-urban areas of Dar es Salaam. Adaptive capacity, crucial to reducing vulnerability to environmental transformations (see Sect. 5.2), is constituted by *autonomous adaptation practices*, which are unplanned and spontaneous, and by *planned adaptation* (institutions) (Stern 2006). The purpose of investigating the synergies and contrasts that may exist between the two modalities of adaptation, and the purpose of identifying the factors that constitute *adaptive capacity*, is to clarify the potential to bridge the gap between formal strategies of spatial planning and management and people's lifestyles in peri-urban areas, which generally depend directly on natural resources and 'informal' management and production of space.

The central question around which the present study is constructed is whether a reinterpretation of sub-Saharan cities through the concept of adaptive capacity to environmental change can contribute to the development of a different interpretive model of the contemporary city. In other words, whether it is possible to build a new interpretation of the contemporary city that challenges and develops alternatives to 'dominant' planning approaches, and overcomes the dichotomies, categories, and partial approaches of traditional urban geographies. In operative terms, the research question seeks to translate the interpretive approach, centred on informal practices and on the agency of people, into the terminology of adaptive capacity in order to bring the issue of environment and environmental changes to the fore in the literature on sub-Saharan Africa.

2.3 Introduction to the Case Study: Peri-urban Dar es Salaam

The development of the case study, from the selection of the city to the fieldwork and data analysis, is constantly linked with the cumulative theoretical approach outlined in the above paragraphs. The following section introduces the salient elements of the present study, conducted over approximately three years (2009–2012) with constant support from Ardhi University in Dar es Salaam. The results are presented in subsequent chapters, together with further theoretical material that became necessary as the research progressed in order to respond to questions raised by the primary data.

This study focuses on the interaction between urban development and environmental change in peri-urban areas of the sub-Saharan city of Dar es Salaam. It seeks to reconstruct the synergetic effects of such interactions on the totality of resources that are available and/or accessible to the households that live in those areas, and to understand how their livelihoods change accordingly. In other words, this research seeks to investigate how the households that currently depend on natural resources (land, water, vegetation, etc.) change their relationship with natural ecosystems, in terms of environmental management practices and resource use, in order to meet their needs and face future and ongoing environmental changes. More specifically, the study inquires as to whether diversification of income sources continues to be people's main strategy for coping with those changes, whether the practice of urban agriculture still maintains a prominent role in such diversifications, and to what extent the combined effect of urban development and climate change condition the sustainability thereof.

These knowledge elements are considered essential, not only in order to shed light on the development and environmental management dynamics in peri-urban areas and their interdependence with urban areas, but also in order to define what conditions are necessary for effective adaptation. In the face of environmental change, what do peri-urban inhabitants count on to secure their future? Do they consider the possibility of using other natural resources? Do they try to achieve greater independence from natural resources by starting non-agricultural activities? Do they embrace a totally urban lifestyle? The answers to such questions have been researched through a series of investigations undertaken in the peri-urban areas of the city of Dar es Salaam in Tanzania.

The choice of Dar es Salaam as a case study was made on the basis of several criteria and preliminary considerations.

First, the research focused on the low- and medium-income countries of sub-Saharan Africa, since they are particularly affected by environmental changes related to climate change, according to the scenarios formulated by the IPCC (2007, 2013). Tanzania is one such country that moreover had already developed programmes and strategies for adaptation to climate change at the national level. In particular, to access the UNFCCC fund for adaptation for *Least Developed Countries* (LDCs), the Tanzanian government presented the *National Adaptation*

Programme of Action (NAPA) in 2007, the National Climate Change Strategy in 2014, and developed several local adaptation and risk reduction plans, whose purpose is to implement the priority projects identified in the NAPA.

While these planning efforts have concentrated mostly on rural regions, the government has also paid attention to Dar es Salaam, and studies to evaluate the city's vulnerability and specific adaptation options have been developed (Dodman et al. 2011; Kebede/Nicholls 2012; other studies elaborated in the CLUVA[21] project and ACCDAR project;[22]). In fact, as a result of its location and physical configuration, Dar es Salaam constantly feels the effects of climate variability, with periodic floods and droughts as well as coastal erosion and progressive salinization of the aquifer (Faldi 2011; Faldi/Rossi 2014). These effects have notable repercussions in terms of economic activity and people's living conditions, in addition to their impact on the environment upon which many inhabitants depend for their livelihoods (UN Habitat 2009).

Furthermore, Dar es Salaam, like many sub-Saharan cities, contains an extensive peri-urban area with 'hybrid' characteristics, contamination of rural and urban forms and practices, and temporary uses, activities, and 'informal' settlements. Such areas have been the subject of numerous studies by Tanzanian academics (Kombe/Kreibich 2000; Kombe 2005; Kironde 2001; Lupala 2002a, b; Davila 2002; Ricci 2012b), and the debate over what future planning for those areas should be is particularly lively, especially following recent discussions of the city's new strategic plan. As regards the objectives of the present research, the absence or inadequacy of planning instruments and policies that consider 'non-urban' settlements and 'informal' activities is an additional reason to develop interpretive models that are capable of nurturing alternative visions and strategies for this part of the city.

This study is built on the assumptions that peri-urban areas and the activities undertaken therein are an integral part of the city and play an absolutely relevant role in urbanization processes, and that the phenomena and characteristics that affect them must be included in the planning process (and in environmental management) as a fundamental resource. The research hypothesis is that peri-urban areas' adaptive capacity to climate change depends on four main factors:

1. The typology and totality of the environmental impacts of climate change at the local level. *What environmental cycles will be modified? To what extent?*
2. Rural–urban dynamics and relationships, land use to the urban fabric. *If and how are they/will they be impacted by the environmental transformations caused by*

[21]CLUVA is a Seven Framework Programme project which aims to develop methods and knowledge for African cities to manage climate risks, to reduce vulnerabilities, and to improve their coping capacity and resilience towards climate changes. It focuses on five African cities, including Dar es Salaam. Further information about the project is available at: http://www.cluva.eu/.

[22]*Adapting to Climate Change in Coastal Dar es Salaam* (ACCDAR) is a three-year project co-funded by the European Commission. Most of the studies on biophysical and social vulnerability in Dar es Salaam are accessible at: http://www.planning4adaptation.eu/041_Papers.aspx.

climate change? How and to what extent do they contribute to the resilience of the urban and regional systems to which they belong? How does climate change impact urbanization processes by accelerating or slowing anticipated dynamics?

3. Autonomous local capacity to address the consequences of climate change. *Is there knowledge or experience related to adaptation to environmental change and climate variability? On what key factors and local actors is adaptive capacity based?*
4. Institutional capacity for environmental management and urban planning. *Does a planning system exist for peri-urban areas? At what level of government? How effective is it, and for which issues? Do local administrations see funds for adaptation to climate change as an opportunity? If so, to do what?*

The field research included four types of investigation. The first is Household Questionnaires, which sought to identify lifestyles and modalities of resource use in peri-urban areas (30 questionnaires were administered in 6 subwards of peri-urban areas, and 10 in 2 subwards of an urban area). The questionnaire drafting and the face-to-face administration process played an important role in the collection of information relative to the various aspects investigated, but above all it allowed for a continuous learning process through which different types of language and knowledge were tested. While organizing the research and questionnaire administration, it was possible to observe and experience different areas of the city, and to discover the relationships between the inhabitants of peri-urban areas and formal or informal institutions, as well as the roles of and interactions between such institutions. Although such information was not the direct result of the household questionnaires, it constitutes an equally, if not more important contribution to knowledge of peri-urban areas and their dynamics.

The second type of investigation involved Ward Questionnaires (WEO—*Ward Executive Officer*, SWO—*Subward Officer*) and District Questionnaires (WPO—Planning Department, Environmental Engineering Department (water management), Agriculture and Livestock Department). These included a series of interviews and questionnaires conducted in various institutional sectors at the local level, in order to gain an understanding of policies, urban planning, and environmental management instruments in peri-urban areas, and projects that were either already concluded, under way or in the planning stages.

The kind of research entailed field surveys and database reviews aimed at gathering data and information on the actual state of natural resources, infrastructure and services, land use and informal activities, and environmental pressures and criticalities in the peri-urban sphere.

Lastly, interviews with national level research centres, governmental institutions, and NGOS were conducted in order to understand from a policy perspective the programmes and instruments used to manage natural resources, adapt to climate change, and implement the NAPA, national policies on urban and territorial development, and environmental and urban planning strategies at the national and local level.

Moreover, the results of this field study have been validated and enriched with findings from successive investigations conducted between 2010 and 2013 under the Adaptation to Climate Change in Coastal Dar es Salaam (ACCDAR) project.[23] Together with multidisciplinary investigation of biophysical and social vulnerability to environmental change, the ACCDAR project also included the administration of the above-mentioned household questionnaire to 6000 households in peri-urban areas of Dar es Salaam located in the coastal plain.

2.3.1 Household Questionnaires

This questionnaire was designed to collect information that would be useful in terms of defining the elements and dynamics through which the ‘platforms’ of action (to use Simone’s terminology) that allow people to address environmental change are constructed. In other words, the goal was to understand how such platforms are constructed, how they operate in order to reproduce life in peri-urban areas, and how they allow people to connect social transactions and livelihoods in order to adapt to or resist the changes underway (elements that will be defined as autonomous adaptive capacity).

The questionnaire’s four areas of inquiry were defined on the basis of this perspective.

Households were chosen as the universe of study because this allows for a better understanding of how and why people organize their collective and individual processes and activities within the family’s decision-making mechanism (Preston 1994). The livelihood strategies of the family nucleus therefore provide relevant information in terms of collective and individual processes of environmental management and adaptation strategies.

The specific objective of the household questionnaires was to collect information on the following four areas of investigation, which were identified through a review of the literature on peri-urban areas (aimed at identifying the elements needed to define point 3 of the research hypothesis, regarding the context being studied and discussed in the previous paragraph):

I. Rural–Urban Interaction
II. Access to environmental resources and services (land, water, energy, etc.)
III. Management of resources (or environment) (water, waste, land, etc.)
IV. Climate Change: environmental changes and autonomous adaptation strategies.

[23]Adaptation to Climate Change in Coastal Dar es Salaam is a three-year project co-funded by the European commission. More information about the project is available at: http://www.planning4adaptation.eu/.

Rural–urban interactions, the economic flow of resources and socio-cultural relationships is fundamental to understanding the dynamics of urban and territorial development. The rural–urban dichotomy has been addressed in key documents that are at the crux of the UN Habitat mandate. As emphasized by the Istanbul Declaration and paragraphs 163 and 169 of the Habitat Agenda, urban and rural areas are interdependent from an economic, social, and environmental point of view. An integrated approach and balanced and mutually supported rural–urban development are therefore necessary. As such, the first area of investigation seeks to understand the environmental challenges and opportunities created by the rural–urban interaction (Allen/You 2002), which manifest most obviously in the peri-urban interface and are an essential element for implementing any planning practice. From this perspective, rural–urban interactions also indicate the spatial and economic interdependencies that exist between the two worlds (both are present in peri-urban and urban areas) as elements upon which several autonomous adaptation options are based.

The questions related to *access to resources* aim to identify, on one hand, the use and management regime in the distinct ecological and socio-economic system under the uncertain institutional regime (Allen 2006) that is characteristic of the peri-urban interface. On the other, those questions are geared to obtaining a first analysis of vulnerability and adaptation possibilities in peri-urban areas. The possibility of accessing water, land, the coast, and primary materials constitutes a determinative factor of the adaptive capacity of communities (Satterthwaite 2007).

Similarly, modalities of *resource management* (water, refuse, energy, etc.), which were researched using the third type of data collection, is relevant as regards the vulnerability of resident communities in a given territory. For example, in a context such as Dar es Salaam where solid refuse is often abandoned on the ground, buried, left in waterways or burned due to the fact that there is no refuse collection system, one can predict environmental and health risks related to events such as flooding, strong rains or water provisioning from surface and underground bodies of water that come in contact with dumped refuse. Access to and management of resources reflects how the formal dimension of access to land, water, and other natural resources, which are based on neighbourhood relationships and social networks, are fundamental in determining alternative solutions in the event of changes to certain conditions in the environment (e.g. the drying up of a water source). Where the formal action of institutions is partial or absent, resource management also reflects the importance, if not the necessity of a collaborative dimension in the protection of resources and the environment.

Lastly, the questionnaire was designed to acquire knowledge on the *practices and strategies of adaptation* to environmental change being implemented at the local level, which must be considered when identifying priority adaptation actions, as established at the COP 7 (Decision 28/CP.7). Specifically, the present study sought to determine which environmental changes had been observed by residents of peri-urban areas, to what cause residents attributed those changes, and the strategies they adopted in order to address them in both the medium- and long-term. Autonomous adaptation practices are an example of how the three previous areas of inquiry, combined with other aspects, can lead to decisions aimed at eliminating or mitigating negative impacts of the environmental changes underway, which can alter access to and management of resources as well as livelihoods (whether or not they are linked to income production), thus implicating alterations to rural–urban interdependence.

Variables deemed strategic in terms of adaptive capacity in peri-urban areas were also defined. A sample of households that was considered representative of the local reality was defined and involved in a pilot study that sought to verify the effective importance of those variables and the validity of the questionnaire with respect to the research objectives.

The selection of universe type and sample unit was defined on the basis of spatial and sociological criteria that could be subdivided into three main groups.

2.3.1.1 Selection Criteria for Areas

The areas in which the questionnaires were administered were selected following a series of site visits in the three municipalities of Dar es Salaam and an analysis of the literature on the city's peri-urban areas. Wards were selected that had the following characteristics:

- presence of rural and urban activities (agriculture, animal husbandry, commercial activities, schools, transportation);
- presence of informal settlements and activities;
- medium-low density settlements (one lot equal to 0.08–6.0 ha);
- settlements located in areas with different environmental characters (coastal areas and hinterlands with different morphologies);
- settlements proximate to relevant natural resources (rivers, ocean, humid areas, forests).

2.3.1.2 Selection Criteria for Households

The identification of which households to involve in the questionnaires was conducted with the support of subward officers, community leaders, and street leaders on the basis of the geographical requirements outlined above. Family selection was also based on criteria of socio-economic and cultural heterogeneity in order to obtain a representative sample in terms of income, education, and family composition. In addition, all households selected were dependent on rural and urban activities and resources and settled in the area in a generally stable manner, and thus were in a position to have knowledge of local resources and evolving local dynamics.

2.3.1.3 Questionnaire Design

The questionnaire was prepared with the collaboration and support of the Department of Environmental Engineering at ARU University in Dar es Salaam. An initial draft of the questionnaire was discussed and revised with urban planners and environmental engineers from that university who then approved the final version and the Swahili translation, which made it possible to interview subjects of any cultural level. The final questionnaire is semi-structured and contains 21 questions which are subdivided according to the four thematic sections discussed above. The closed-form questions are multiple choice, hierarchical or dichotomous, and the open questions left space for interviewees to leave comments and suggestions.

The choice of a semi-structured questionnaire as opposed to an open interview facilitated organized data collection and analysis. Within the questionnaire, independent and dependent variables are identified, and through the relationship between them and the physical-environmental variables, information was gathered with which to define the knowledge framework of the pilot study.

The questionnaire administered to the chosen sample is divided into five parts. The first part is related to the interviewee's basic data (age, gender, employment, education, etc.) and the other four parts are related to the four areas of inquiry identified above.

2.3.1.4 Questionnaire Administration

Authorization from the same institutions that approved the fieldwork was necessary for the administration of household questionnaires. A declaration of interest, support and shared research objectives from a local institution (ARU University) was needed, as well as the corresponding request for authorization to carry out such research in the territory of interest. That request was made to the Municipal Director of the Kinondoni municipality, who granted authorization to obtain the consent of WEOs (*Ward Executive Officers*) and Subward Officers in the wards of interest. They then appointed a *Mtaa Executive Officer* (MEO) or a Community Leader to support the organization and administration of questionnaires.

Dar es Salaam is a city-region, and has a regional administration with a Dar es Salaam Regional Commissioner, as well as a City Council led by the Mayor. It is comprised of three Municipal Councils (Ilala, Knondoni and Temeke), which correspond to the three districts of the region of Dar es Salaam. Districts are subdivided into divisions, which are then divided into wards. Wards are divided into subwards, which in turn contain streets (*mitaa*) in the case of urban and peri-urban areas, and villages in the case of rural areas.[24]

[24]Divisions contain many Wards, which are comparable to districts, while *mtaa* and villages are comparable to neighbourhoods.

In light of the criteria outlined above, questionnaires were administered in the Kinondoni District because Ilala, where the economic-operational centre of the city is located, is densely built-up, while the Temeke District is mainly rural due to the near total absence of road networks. Within the Kinondoni district, four wards were selected: Kawe, Kunduchi, Bunju and Msasani (Fig. 2.1). In each ward, two sub-wards were selected and five questionnaires were administered in each. Specifically, in Kawe the subwards of Makongo and Changanikeni were chosen, both of which are in the internal area of the subward rather than the coast, because the coast of Kawe is highly urbanized; in Kunduchi, the subwards of Mtongani, on the coast, and Madale, in the west of the internal zone of the ward, were chosen. In Bunju, on the other hand, questionnaires were administered in Boco, located on the northern coast, and in Bunju A, in the interior of the area. A total of 40 questionnaires were administered, 10 of which were collected in the Msasani Ward, which is completely urbanized and therefore not included in the present study, but which is useful in terms of identifying the differences between urban and peri-urban areas within the themes explored in the questionnaire. In the selected areas, interviewers administered questionnaires in Swahili to the households identified by the Community Leader or *Mtaa* Leader who supported the questionnaire administration, together with personnel from ARU University.

2.3.1.5 Data Analysis

The data collected through the questionnaires was analyzed using both qualitative and quantitative methods. Findings from this analysis where integrated with results from the extensive questionnaire (involving 5 % of the Dar es Salaam population) administered under the ACCDAR project between September and November 2011. The questionnaire involved a broader sample; however, it includes all the wards sampled in the first household questionnaire administered in 2009, which was used as pilot investigation.

The data collected through the questionnaires administered in 2011 were reorganized in order to facilitate analysis using statistical software. Two main types of analyses were carried out.

A first, univariate analysis was conducted using SPSS in order to extrapolate the data through frequency analysis and determine how many times a particular response was given within the study sample, or how many times a given variable had the same value. Absolute frequency was obtained for each variable, which indicates the total number of cases in which a given variable or a modality thereof occurs, and the relative (or valid) frequency, which is based on the relationship between absolute frequency and the total number of cases examined. The relative frequency percentage provides an immediate picture of the situation and allows for rapid comparisons.

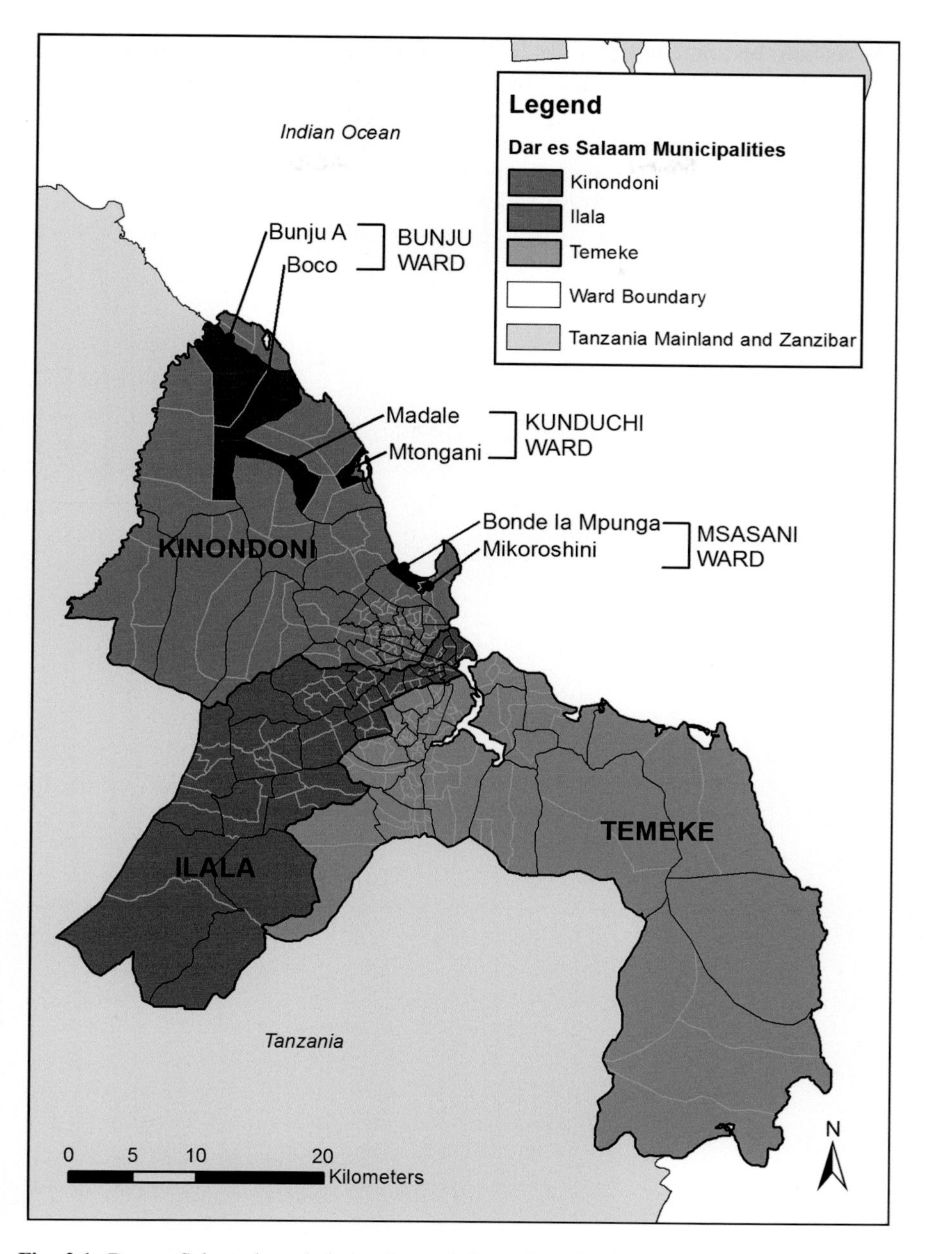

Fig. 2.1 Dar es Salaam boundaries and sampled wards and subwards. *Source* Administrative boundaries (Dar es Salaam City Council 2007)

A second, bivariate analysis was conducted using the distribution of conjoined frequencies, or the intersection of two simple (or univariate) frequency distributions. This facilitated the development of contingency tables in which the relationship between two variables was examined, highlighting the cases in which a concomitant variation of respective variables occurred (e.g. variations in education correspond to an individual's occupation).

Chapter 3
Interpreting the Sub-Saharan City: Approaches for Urban Development

Abstract This chapter reviews the key historical and political points of the urbanization process in sub-Saharan Africa, including the difficult 'modernization' process, promoted first by colonial policies and later by development aid. The distinct characteristics of sub-Saharan cities are then presented, specifically their hybrid rural–urban features, their 'informal' development modality, and their relationship with environmental transformation. The second part of the chapter introduces the case of Dar es Salaam as representative of the cities of sub-Saharan Africa, as well as part of the evidence obtained from the questionnaires or extracted from the literature. The city is analyzed with reference to the evolution of peri-urban areas and the challenges in contemporary planning are discussed, as well as some of the constraints on policies and planning approaches adopted in Dar es Salaam and sub-Saharan Africa generally.

Keywords Modernization · Development · Aid · Peri-urban · Informality · Urban planning · Structural adjustment · Adaptation planning · Tanzania

3.1 The Politics of 'Modernization' and Urban Development in Sub-Saharan Africa

The scientific literature on contemporary African cities has tended to concentrate on characteristics such as the lack of regulation and the chaotic nature of urban development (Rakodi 2002). This portrait of African cities as places that have fallen into chaos as a result of the collapse of effective governance, without infrastructure or services, derives in part from an uncritical acceptance of a normative ideal that is linked to the idea of the 'good cities' of Europe and North America. The general tendency has been to respond to the perception of uncontrolled urban growth by imposing a minimum of 'order' on the city (Murray/Myers 2006: 17).

Following the waves of urban neoliberal reforms promoted elsewhere, African cities have undergone a profound transformation over the course of the last few

L. Ricci, *Reinterpreting Sub-Saharan Cities through the Concept of Adaptive Capacity*, SpringerBriefs in Environment, Security, Development and Peace 26, DOI 10.1007/978-3-319-27126-2_3

decades, in terms of the form and the function of local government. This transition from the old style of local government action (characterized by formal bureaucratic and hierarchical organization and the desire to control) to new forms of urban governance has led to the introduction of political initiatives based on economic competition and greater participation of groups of enterprises in local decision-making processes. The adoption of new urban governance modalities has led to a significant reduction of municipal functions and responsibilities (Murray/Myers 2006). Critical approaches to these policies maintain that the implementation of 'good governance' measures often fails to lead to democratization, causing instead a loss of power among local bodies and communities. For example, Beall et al. (2002: 65) assert that in the hands of neoliberal urbanists, 'governance' is an empty abstraction, a rhetorical device that effectively reproduces what is defined as exclusive democracy in African cities (Abrahamsen 2000, in Murray/Myers 2006: 19). Exclusive democracies "allow for political competition but cannot incorporate or respond to the demands of the majority in any meaningful way" (Abrahamsen 2000: xiv).

The result is that the contemporary African city seems to have gone ahead on its own, beyond any planning scheme. Through what is defined as an informal process of urbanization, space is produced according to unplanned and unforeseeable modalities that nevertheless interweave with formal local systems and the global market through relationships that are often characterized by power imbalances and social injustice.

3.1.1 From Colonialism to the Politics of Development Aid: Urbanization and Planning Approaches

The following parenthesis on the evolution of the African city seeks to give an overview of the variety of activities that have made urban life possible, and of the social forms that Africans have created. Particular emphasis will be placed on the manner in which planning policies and strategies have determined the characters of cities, and the legacy they have left for the contemporary African city.

Although relatively obvious vestiges still remain from the cities of the past (Pre-classical, Classical and Islamic), the form and characteristics that African cities have assumed today most clearly reflect signs of the colonial period, during which the foundation and growth of the city was linked to commerce, and urban development was principally based on exploitation (Freund 2007: 65).

Urban expansion proceeded under the colonial occupation of the late nineteenth century, but it intensified in the last phase of the colonial regime following the Second World War. It was during this period that the urban population increased considerably; from 1920 to 1960, the percentage of people residing in urban areas grew from 4.8 to 14.2 (Benatia 1980, in Freund 2007: 65). This growth in the population and the dimensions of the city reflected the intensification of commercial vocations in African cities. Nevertheless, not all urban expansion was linked to commercial expansion; some cities developed around production sites, as was the case for the Copper belt cities between the Congo and Zambia. Moreover, for some cities colonization brought decline rather than expansion, as was the case for the cities located on old caravan routes that were cut out by newer commercial routes oriented to exporting primary materials towards the ports of colonized countries. Sociological theories on 'parasitic Cities' often refer to this phenomenon (mentioned in Chap. 2), whose welfare was based on resources arriving from rural areas and on impoverished farmers. In the main cities of the colonial period there were, therefore, new forms of well-being based on social control and exploitation.

On the one hand, colonialism allowed for the formation of a plural society, in which culturally defined groups that were apparently separate and antagonistic met in the market place and were able to maintain peaceful relations while under colonial protection. However, on the other hand, it imposed severe segregation between natives and non-natives, which was justified in part by cultural imperatives and "sanitary syndromes" (Freund 2007: 717–779) which claimed that segregation reduced the risk of disease. These principles guided the construction and reconstruction of many African cities, as well as the modernist and post-modernists policies mentioned above.

3.1.1.1 The Urban Policies of Colonialism

During the modern colonial period, which began in the nineteenth century, in some cases, cities were developed *ex novo* and even maintained continuity and very important connections with rural life. In other cases, pre-colonial cities sprang up that were only marginally modified by new colonial structures (Freund 2007: 101). Colonial domination, characterized by functionalist political economy and by notions of racial segregation, often sought to impose idealized norms and forms of the metropolis, but the African reality continued to find space within these attempts. For this reason, and despite the fact that they have also gone through periods of stability, colonial cities have often experienced social and political crises that have revealed the contradictions of colonial development. Throughout the many phases of the history of urban Africa, one finds examples of cities that have maintained characteristics that were culturally and physico-spatially African, such as the coexistence of rural–urban activities and forms (Freund 2007: 65).

The schools of English and French sociology have developed two different interpretive approaches to the urbanization processes of the colonial period. While the English schools of thought (and perhaps those of today as well) were focused on how much of the old African culture had survived the hard life of cities and theorized a 'new indigenous civilization', the French school, with the work of Georges Balandier (1965), laid the foundations for a reflection that sees Africans as citizens who may potentially reclaim their 'own' cities, which would be different from an apparently traditional and concluded settlement, conserved by the settlers with good taste (Coquery-Vidrovitch 2005: xvi).

During the colonial period, the policies oriented to meeting the needs of the urban elite were already becoming less relevant. Several colonial authorities (e.g.

Belgian) recognized the existence of an urban population that predated colonial control and performed a useful economic function. Although colonial planning at no point conceived of the birth of a large city characterized by the presence of an 'African proletariat', nor did it envisage the need for cultural and economic continuity with the past, planning did seek to reduce racial segregation around the 1950s (with the exceptions of South Africa and neighbouring territories) and began to apply modernist ideas based on the social, democratic and liberal ideals of contemporary Europe, which were sometimes opportune and sometimes not (Freund 2007: 98).

The colonization period in Africa was an almost uninterrupted period of centralized administrative rule. The first decentralization period occurred from the late 1940s through the early 1960s. Local (and state) authorities were established by mutual agreement between the nationalists and the departing colonial authorities, and attempts were made at formalizing existing understandings about the democratic delivery of local services (which closely resembled local governments and communes). While to the British, local government was a 'school for democracy', for the French the commune was the place of civic rights for both urban and rural citizens (Stren/Eyoh 2007).

In the mid-twentieth century, and following the trauma of the Second World War, a new international order emerged that, despite having its origins in previous centuries, explicitly reflected a desire to improve international economic collaboration and reduce conflict between nations. To that end, a series of multilateral institutions was created (the United Nations, the International Monetary Fund, the World Bank, etc.) that initially concentrated on the reconstruction of Europe with the Marshall Plan, and subsequently pushed for the decolonization of Africa and other colonial territories. This historic development is marked by Truman's speech on January 20, 1949, in which he launched the development and progress phase, with important consequences for the urban policies of the Global North and South.

During the 1950s and the 1960s, a theory of modernization prevailed that emphasized processes of industrial expansion guided by national states in order to accelerate economic growth and poverty reduction. Following this line of reasoning, colonial governments tended to create modern structures of political representation and cities grew in terms of both dimensions and importance. Within such structures, African politicians, for whom the urban arena had become very attractive, slowly began to find a space, to acquire credit, power and the ability to control the resources with which to jump-start the 'urban political machine', and to offer symbols of 'progress' to the electorate through provincial and territorial self-governance (Freund 2007: 100–101). However, they did not attempt in any real way to confront the difficult and contradictory questions posed by the planning of a modern city and by the mass urbanization of the late colonial period, an undertaking that was rendered even more difficult by the great variety of urban populations in colonial Africa. It was not only a question of racial minorities, but also of the differences between natives and migrants, owners and renters, traditional chiefs and the educated elite. The challenge faced by city planning therefore consisted of many different aspects.

3.1.1.2 Urban Policies of Early Independence

The end of the colonial period generally coincides with 1960, when the majority of African nations achieved independence (Nigeria, Congo and the French colonies of Sub-Saharan Africa, which were preceded by Ghana, Sudan and the countries of North Africa, with the exception of Algeria), though the process was concluded only in 1990, when the last colonial state, Namibia, was declared independent. The era of independence was a period of global change that had notable impacts on urban development. In some cases, independence was marked by a restructuring of the largest old colonial cities, which in turn had some influence urban growth. The importance of the old capitals of Western Africa declined, and the 'need' for national capitals grew, a need that led to a radical renewal of several cities (Gaborone in Botswana, Kigali in Rwanda, and Nouakchott in Mauritania) or to the foundation of new capitals (Abuja in Nigeria, Dodoma in Tanzania and Lilongwe in Malawi). The role of city-leader became increasingly visible, there was a considerable growth of administrative centres and the city became a place of modernity where Africans could become cosmopolitan, where there were institutions of higher education, development agencies, sports stadiums, hotels, etc. A good quality of life was theoretically guaranteed for everyone and racial restrictions were eliminated (though perhaps only formally).

Power was generally in the hands of the central government, which sent its delegates to supplant local authorities; national ministers controlled the city and the correct administration of a city depended on their generosity. Independence did not represent a true interruption of city administration during the colonial period. In the early years, the character of planning and the economic structures of the late colonial period persisted, even if the white managers had become transitional (rather than colonizers). In fact, in many instances, the old elite did not immediately disappear, and the white population actually increased after independence (e.g. Abidjan), a sign that much of the bureaucracy yielded to the pressure for continuity, though there were exceptions, notably in North Africa (Freund 2007: 146).

With independence and rapid urbanization of the rural population, the criticality of modernist development also emerged, and many cities sank into decline and extreme poverty. In fact, the urban African population at the beginning of the 1970s had grown so much that the local economy could not absorb it (see 'over-urbanization' in Sect. 1.1.2). The colonial phobia of concentrating large numbers of impoverished and semi-educated people took on a new form. Such individuals were seen as social parasites, which thronged the agrarian and mining export-production zones of rural Africa in order to settle in swarming shantytowns. Beginning in the 1960s, the majority of African countries, irrespective of political inclination, organized systematic raids and expulsions of the urban population that lived in self-constructed poor settlements.[1]

[1] One of the last of this kind of raid was described by Aili Mari Tripp with reference to the 1983 Nguvu Kazi (power labour) campaign in Dar es Salaam. The state tried to define a large number of

3.1.1.3 Urban Policies in the Crisis Era

In the 1970s, with the political agitation, criticisms of modernization and power imbalances and the development of *dependency theory*, the emphasis on production and growth shifted to redistribution and poverty reduction through policies that supported production to meet basic needs by improving participation and empowerment of the population.

At the end of the 1970s, the modernist Africa that had prospered immediately after independence led to crisis in major cities (beginning with Lagos and Kinshasa) due the combined effect of several disastrous circumstances (limited public transport, chaotic land use regimes, difficult access to structured and remunerated work, lack of public space that, according to conventional modernist jargon, would have given citizens a sense of belonging and pride, etc.), and became representative of a new set of city planning problems. As such, the uncontrolled spreading of large-scale settlements swept away the functional city planning that had been inherited from the colonial period, when construction of shanties seemed to be more or less under control (Freund 2007: 150). With the crisis blocking modernist development, development experts, who were often the principle critical observers of African cities, proposed a new approach. Parasites, inhabitants of shantytowns, and unemployed women began to be viewed as the authentic constructors of African cities as a part of grass-roots development processes. A study by Andrew Hake on Nairobi—the 'self-help' city, as he called it—published in 1974, constitutes yet another emblematic case of that process: "Far from being parasites, such poor dwellers in the city are there for a reason, to make themselves and their households a better life; they perform important services, create their own employment and make useful contributions to the economy. Such people, far from dragging down the economy, are actually engaged in building it up" (Freund 2007: 154–156).

These approaches certainly represent a fundamental turning point, but they still viewed the role of inhabitants as a function of the imperative of economic growth that continued to guide the development of cities.

In this new perspective, the belief developed among Western donors and scholars that the African state had become overcrowded and corrupt, that it was more of a nuisance and a burden than a real contributor to development. The state announced planning rules and regulations that were little different from colonial models, except for the fact they no longer had the capacity and resources to

(Footnote 1 continued)

the city's inhabitants as "unproductive" (including shoe shiners and women who could not officially be classified as married). This vision has never been accepted as "legitimate" by the population, rather it is considered "intended purely to relegate even more economic activities to the realm of the illegal until, as was typical, the authorities got tired and abandoned the campaign" (Freund 2007: 149).

implement their plans. Those plans were in certain respects detrimental insofar as they favoured certain social groups, and in other respects they were irrelevant to the real social processes underway in the city.

In the late 1970s and early 1980s, a second phase of decentralization characterized sub-Saharan Africa, when in response to post-independence over-centralization and the problems of local project implementation, central governments relocated development committees, technical ministries and large projects to the district level (e.g. Nyerere in Tanzania 1972). However, expectations for this second wave of decentralization were not fulfilled due to the failures of large projects and the absence of finance. The beginning of the structural adjustment period of the mid-1980s concluded this phase of decentralization, although a third phase involving non-state actors occurred at the end of the twentieth century.

3.1.1.4 The City in the Era of Structural Adjustments

During the 1980s and 1990s, the development of 'neo-liberal' theories, and the implementation of Structural Adjustment Programmes oriented urban development policies towards privatization, liberalization and deregulation. They led to a push for the reduction of state power in light of the destructive impact they had on the market and on people's lives. As a result, efforts were made to severely limit transfers to urban areas, except when they were meant to meet the needs of the elite. The areas at the edges of cities, on the other hand, became the subject of plans to parcel the land, which were seen as an economical and adequate method of allowing for urban growth. These plans applied to green, undeveloped areas and provided for infrastructure that would guarantee a minimum level of services, assuming that the new residents would build whatever habitation they wanted, however and whenever they could, according to the self-help ideology. But the application of that strategy proved to be too costly, and the necessary infrastructure either was not developed or was subject to very limited maintenance, and as a result a *laissez-faire* approach ultimately prevailed.

The result of these policies was both a reduction in rural–urban migration and an increase in construction within the city. In fact, liberalization of the financial market pushed many Africans living abroad to send money to Africa, while liberalization of commerce allowed for the importation of a variety of materials, including construction materials and vehicles that played an important role in the expansion of the city. Moreover, the reduction of state participation in the economy through privatization and deregulation led to an increase in investments, but they focused more on consumption than production. The investments in the industrial plants of Latin America were all but absent in Africa. A majority of investments concentrated, rather, on the housing sector, favouring expansion of the city, particularly in peri-urban areas (Briggs/Yeboaht 2001: 19–21). Single builders were ready to build in non-urbanized areas, anticipating the future provision of services and infrastructure and benefitting from the lower price of land in such areas.

3.1.1.5 The Policies for Urban 'Good Governance'

At the beginning of the twenty-first century, it began to be clear that the development policies implemented over the previous decade had not yielded the benefits they had promised. Deborah Potts, sceptical of the advantages of leaving urban life in the hands of the market, highlighted the savagery of the anti-urban politics of the Structural Adjustment Programmes directed at people already impoverished and under siege. She underlines the fact that many Africans do live in cities, and that African cities continue to grow through complex forms of negotiation, which at the administrative level mix the modern normative dimension with the local reality of power legitimization (Freund 2007: 156–157).

Contemporary urban development is mainly concentrated on governance and incentives for structures and relations that sustain states and markets (Beall/Forx 2009: 11).

The concept of good governance, intended to overcome conventional forms of state, is heavily promoted by international agencies. The State itself is called on to define autonomous organisms that are more in touch with reality and more sensitive to people. A classic manoeuvre is the imitation of French decentralization in Francophone countries, with the subdivision of large municipalities into smaller units that are governed by administrations elected directly by the people. In other cases, development agencies have pursued improvements in certain areas by concentrating on specific activities, promoting what has come to be defined as the local economic development.

By the late 1980s and early 1990s, a great number of states undertook decentralization measures from the centre to local governments, with the support of civil society. This third phase of decentralization, (not yet completed) is characterized by the growth of civil society (particularly in urban areas), and by democratization and re-democratization that began in the 1980s. These two trends have country-specific shapes and dynamics, and although they are independent of decentralization, they do contribute to reinforcing it. Unlike the earlier decentralization initiatives, this most recent phase has received more social and political support, contributing to endogenous rather than externally induced decentralization actions (Stren/Eyoh 2007).

The lack of funding and the lack of local authorities' capacity to raise their own funds often structures the ties of dependency between associations, financial sources and lobbies, which invest in local development. Non-Governmental Organisations and *Community-Based Organizations* (CBOs) that are able to mobilize a large number of people to take on an important role in this context. CBOs represent vital networks with pre-colonial roots that have multiplied and taken on different forms over time. They are often attributed with having improved urban life in Africa, thanks to examples of Community Development Organizations that have successfully managed and maintained the urban environment.

It is not clear whether CBOs are organizations that act from the bottom-up, with political conscience and the capacity to acquire a decision-making role in urban development strategies. As regards the basic needs of poor people, Jean-Luc

Piermay, discussing Kinshasa, highlights the fact that "survivalism is not the same as building a real urban social fabric" (or urban development). Moreover, African 'communities' are not necessarily "so welcoming or egalitarian" as the concept usually implies. In the bosom of self-help there lies also exploitation and growing forms of differentiation. It is, moreover, not only the poor who help themselves in situations where the state weakens. Particularly, in those African cities with more substantial middle classes, the possibilities are greater both for new forms of accumulation through uncontrolled urban expansion and for yet greater problems in terms of achieving some kind of sustainable urban planning, as the North African literature indicates" (Freund 2007: 162).

Nevertheless, despite decades of accusations from the West, the central government still plays a very important role in determining the development of the city and in many African cities decentralization and devolution process are still ongoing. One reason could be that the decentralization policy was not accompanied by fiscal decentralization. However, it contributes the mobilization of financial resources for the improvement of infrastructure and services in order to address the needs of a rapidly changing society and the demands of good governance and modernization. From this perspective, the policy of decentralization has been increasingly identified as a function of improved understanding of good practices (Stren/Eyoh 2007).

3.1.1.6 Tendencies of Contemporary Urban Policies

The post-colonial decades added an important level to the history and evolution of the African city. Catastrophist approaches, approaches that exalt local creativity, approaches of good governance or effective government, have been developed in recent years based on a variety of ideas of the African city. Given the failure of the restrictive planning that was in line with the colonial past, much of debate within the discipline now focuses on the question of whether African cities are moving towards new governance regimes and more realistic goals.

Considering the various tendencies at play, to underline the positive aspects of African cities, their creative and original responses to change (global and local) and their possibilities for reform, seems today to be more useful than simply applying a polyvalent 'Afropessimism'. Unfortunately, these interpretive approaches—described in part in Chap. 1—have had very little effect on development in African cities, unlike approaches combining 'security' concerns with the goal of improving the functioning of the State and the Market with an emphasis on the role of property law norms, bureaucratic efficiency and responsibility, which have had a heavy influence. Above all, with the emergence of the spectre of terrorism (the attacks of September 11, 2001), poverty and inequality have come to be viewed as threats to global security and as linked to the failure of medium- and low-income states (Marvin/Hodson 2009). Recent conflicts in the Middle East and North Africa render the question of security, which combines political and environmental aspects, more relevant. That question is further amplified by the debates on Climate Change and the 'environmental crisis' of recent years, which will continue to exacerbate conditions of poverty and decline in both urban and rural areas, and will have worse effects in African cities, particularly in peri-urban areas, where a large portion of the population is concentrated, and on people who are dependent on natural resources.

Although African economies have improved recently, Africa is still the continent where the biggest economic, environmental and urban challenges lie. Many African governments believe that rapid urbanization makes things worse in terms of economic gains (and development). To contrast this growth, three quarters of African governments are implementing policies to reduce rural–urban migration (UN DESA 2012). In recent decades, international organizations are reconsidering the city as an 'engine of growth' and the connection between urbanization and economic growth is considered automatic and inevitable (Turok 2014). To better govern this economic growth and ecological resilience, effective and transparent policies are needed as well as effective national planning and investment linking complementary and differentiated rural and urban strategies (Parnell/Simon 2014).

City Development Strategies (CDSs) have developed and are continuing to develop city-level regional strategic processes; however, they have a donor-driven origin in the neoliberal devolution of planning and marketing roles, which tend to have local legitimacy and represent an opportunity for a bottom-up urban change. To address the challenges associated with the demographic transition, current urban development strategies are mainly focused in ecological vulnerabilities, large infrastructure and land use management reforms. At the urban policy level, legal and institutional reforms are oriented to finding the appropriate size and shape of governments, while tax reform is considered necessary to ensuring linkages between resources and responsibilities (Parnell/Simon 2014). Associated with those strategies are both new and old urban development actions, ranging from settlement upgrading, development of new settlements and formalization of land propriety, etc.

3.2 Urbanization Process in the Cities of Sub-Saharan Africa

Since the medieval Islamic period, Africa has been involved in a global system of commerce between the Indian Ocean, the Mediterranean and the Atlantic, which favoured both autonomous urban development and that which was imposed from the outside. The historical study of urban and territorial transformations in Africa as a discipline in Western universities has evolved from one that was concentrated on a few themes and certain urban centres, to a much broader approach that explores the actions and interactions both within and between specific urban contexts. According to this approach, the complexity of the urban environment cannot be analyzed without including the analysis of peri-urban and non-urban areas (Coquery-Vidrovitch cited in Faloa/Salm 2005: xv).

The contemporary representation of the African city as a place where planning strategies and building codes are either absent or useless, where spatial models of urban growth seem random and the urban landscape appears without any real coherence or general design, is contrasted by the city's capacity to function despite the apparent lack of planning and coherence (as demonstrated by Rem Koolhass 2002 in his study on Lagos) and the self-organization of urban life that occurs in a context of persistent and continuous distress and poverty (Gandy 2006).

But differences aside, the various academics who have studied sub-Saharan cities agree that, notwithstanding local particularities, one of their principle characteristics is the constant presence of vast hybrid rural–urban settlements around the central nucleus. A more detailed investigation of the development modalities in these areas allows for the identification of two other crucial aspects: informal development and its close relationship between the environment and natural resources as strategic components of people's lives. The areas with hybrid rural–urban characters have mostly developed (and continue to do so) in an informal way and in contrast with existing plans, even if their expansion was generally supported by various political or speculative economic interests. This situates those areas in a complex economic and social system that often remains marginal to 'formal' construction processes of the city (see Sect. 4.2.2). Informal economies and settlements condition and are conditioned by environmental characteristics: activities such as urban agriculture, or 'non-conventional'[2] access to water resources, or self-construction of shelters, are strongly dependent on climate and environment (see Sects. 4.2.1 and 4.2.3).

3.2.1 Evolution of Peri-urban Spaces and Hybrid Forms

We have seen how rapid urban growth (Sect. 3.2.2) favours the proliferation of unplanned settlements, and in fact, in recent decades informal peri-urban areas have been the site of the majority of demographic growth in African cities. These processes have moulded a fragmented and dynamic rural–urban interface, characterized by a constant evolution of land uses, activities and social and institutional organizations which has been defined by Simon (2008: 171) as "forms of hybridity" characterized by the coexistence of rural and urban characteristics.

The proliferation of hybrid settlements has been oriented and facilitated by urban and rural development policies. In the colonial period, the creation of racially segregated areas as well as the allocation of land for Africans outside municipal boundaries in rural areas (under traditional, communal or customary tenure arrangements) was associated with peri-urban settlements where indigenous people constructed their livelihood as a function of the colonial elite.

In the early independence period, large plots of rural land, close to the urban settlement, were assigned to farmers and migrants to prepare the areas for urban expansion. The differential rights of access to land associated with the customary, private and state (or municipal) land also contributed to the creation of peri-urban

[2]Informal development implicates the absence of infrastructure and services, and pushes people to search for alternative and diversified options for resource provisioning.

settlements outside formal regulatory control. Land for housing and other basic needs could be obtained easily and cheaply in customary land close to urban areas. For this reason, although such areas generally lacked plans and infrastructure, the upgrading programmes implemented under the neo-liberal policies contributed to the relocation of residents from the upgraded settlements to new and bigger plots on the edge of the city, thus contributing to the expansion of peri-urban settlements.

Although African cities may have different features, they have been characterized since antiquity by a mix of functions, forms and cultures. In Western Africa, during the pre-colonial period, the inhabitants of cities maintained rural dwellings (Freund 2007: 50). Studies on the first large settlements, located in present-day Botswana, define them as agro-towns. Anthropologists have described the Yoruba settlements, in present-day South-Western Nigeria, as already urban prior to the nineteenth century, refuting the simple categorization of a continuous shift from rural to urban, and recognizing a physical and cultural combination of rural and urban (Freund 2007: 3). This hybrid character is also recognized in the settlements of the Ga people in Accra, Ghana. Studies on the colonial period maintain that urbanization should be considered part of a "folk-urban continuum" (Southall, ed. 1973, in Freund 2007: 85), and that tens of thousands of 'new townsmen and townswomen' in Africa were what sociologists later referred to as 'straddlers', or people who live between the city and the country and take advantage of the opportunities that both have to offer wherever possible, in one or both contexts. Similar examples, though with different characteristics and forms, can be found in the literature on urban African history of the subsequent periods. Such a historically and spatially broad series of African settlements, which cannot be defined with the cognitive and theoretical instruments of the main schools of Western urban studies, clearly leads to questions as to what was and is urban in this context. The purpose of this paragraph, and of this research in general, is not to answer this question, but to investigate hybrid forms of settlement and to identify their specific potential.

3.2.1.1 Physical–Environmental Hybridity

Population, housing density, infrastructural characteristics, administrative boundaries and predominant economic activities are the main variables that are conventionally used to distinguish rural from urban (Tacoli 1998). The peri-urban interface has often been defined as a complex phenomenon, often characterized simultaneously by the loss of 'rural' aspects (loss of fertile soil, agricultural land, natural landscape, etc.) and by the lack of 'urban' attributes (low density, lack of accessibility, lack of services and infrastructure, etc.) (Allen 2003) (Fig. 3.1). As previously discussed (1.1.3), attempts to define local and territorial development in this context waver, in many cases, between the desire to emphasize the connections between rural and urban, interpreting such areas as the temporary result of spontaneous processes that rapidly transform rural territories into urban ones, and reference to the concept of 'peri-urban', understood as a term that describes areas with a mix of rural and urban characteristics (Iaquinta/Drescher 2001).

Fig. 3.1 Peri-urban areas in Dar es Salaam, Ilala Municipality. *Source* Ricci (2014)

3.2.1.2 Ecosystemic Hybridity

As underlined in Sect. 3.2.1, an environmental conceptualization that characterizes peri-urban areas as a combination of heterogeneous ecosystems that are 'natural', 'productive', 'agricultural' and 'urban', influenced by flows of materials and energy, and generated by urban and rural systems that condition each other (Allen et al. 1999: 7), has important implications for intervention policies. First, it leads to a broader understanding of the processes under way, drawing attention to social and economic organization, and to biophysical aspects. For example, such a conceptualization can reveal that the acquisition processes of private property can favour an unequal distribution of conditions of environmental quality, a consequence of the combination of building speculation and social segregation. As such, the areas subject to environmental risks often become the living space for the poorest individuals, while the areas of higher environmental quality constitute the epicentre of speculative mechanisms, preventing or limiting less wealthy inhabitants' access for the purpose of productive activities, or damaging the precious ecological processes of natural systems. Moreover, some authors argue that the combination of territorial characteristics, such as carrying capacity, soil productivity, vulnerability to flooding, availability of drinking water, etc., constitutes a more appropriate group of criteria for an environmental analysis of the peri-urban interface with respect to conventional zoning based on density, morphology and rural or urban uses of the territory (Allen 2003: 137–138).[3] This continued interaction and interdependence, linked to commerce and the flows of ecological goods and services, creates the need for reciprocal and sustainable relationships between urban, peri-urban and rural systems, and has led to a reconsideration of the urban system as an environmental one.[4]

3.2.1.3 Social Hybridity

In addition, the presence in peri-urban areas of a highly heterogeneous social composition that is subject to high temporal and spatial dynamism must also be considered. Small-scale farmers, 'informal' residents, industrial entrepreneurs and middle-class 'urban' commuters can all coexist in the same environment, but with interests, practices and perceptions that differ and often conflict with each other. The composition of and interests within these various groups tend to change over time through a process characterized by the incorporation of new variables and stakeholders (Allen 2003).

3.2.1.4 Institutional Hybridity

On the other hand, from the institutional perspective, the peri-urban interface is characterized by the convergence of sectors and the overlap of local bodies with different spatial and physical competencies. This is linked to the changing geographical position of the peri-urban interface, or to the process for which institutional jurisdictions, or zones of responsibility, tend to be either too narrow or too broad, too urban or too rural when addressing questions related to sustainability and poverty (Mattingly 1999: 4–5).

Moreover, even actors in the private sector, non-governmental and local community organizations often intervene in the management of peri-urban areas without a clear direction from governmental structures. The problem of institutional fragmentation is particularly important to understanding the 'limits' faced by environmental planning and management. The coexistence of different administrative units in the same area, the dearth and inadequacy of connections between them, and the limited nature of municipal power in sectors such as transportation, water,

---

[3] "Conventional urban planning has favoured a centrifugal view inadequate for addressing the characteristics of the interface's 'patchwork' structure" (Allen 2003: 137).

[4] One example of this approach is defined in the literature as the 'urban bio-region' (Atkinson 1992).

management of energy, solid and liquid waste and territorial planning create ambiguity as regards which institution is responsible, in which specific zone and what type of intervention is to be planned and/or implemented (Durand-Lasserve 1998). This means that the conceptualization and methodology for environmental planning and management of the peri-urban interface needs to move away from a physical definition of rural and urban zones (conceived as a clear limit of geographical and administrative entities) and towards a broader understanding according to which complex models of settlement and resource use, and the flows of natural resources, capital, goods, services and people are not adapted and assimilated to jurisdictional limits (Allen 2006).

3.2.2 *Rapid Growth and Informal, Unplanned 'Modes of Urbanization'*

The rapid, informal and unplanned growth of sub-Saharan cities represents for some a veritable '*mode* of urbanization' with "an organizing logic, a system of norms that governs the process of urban transformation itself" (Roy 2005: 148), a modality of "production of space" (Roy 2009: 825–827). The peri-urban interface represents the site of informal development, not only in Africa but also in other low- and medium-income countries, where the peri-urban interface is constituted by various forms of informality, including the flows of labour and typologies of habitation at the base of urban life and economy (Breman 2003, cited in Roy 2005: 149).

3.2.2.1 The Production of Informal Settlements

One of the reasons for the high number of informal dwellings is the fact that the production of formal settlements is unable to keep up with population growth, partially because of the scarcity of resources. While new settlements are developed, those that already exist continue to grow. The most repressive norms for containing informal settlements were abolished when political independence was achieved, while the subsequent zoning norms, building codes and many other planning rules continue to exist only on paper in many countries. The rules imposed by urban planning often limited the possibility of providing housing at accessible prices, thus favouring the diffusion of unauthorized or illegal settlements (Hansen/Vaa 2002: 9).

It is nevertheless difficult to establish a distinction between the legality and illegality of such settlements. There are notable differences between juridical contexts of African countries, and as regards the acceptance or recognition of unauthorized settlements. 'Illegal' or 'extra-legal' settlements occur in three main forms: illegal occupation of land that violates collective or individual property rights, parcelling of land in a manner that conflicts with planning rules, and the construction or use of houses without authorization or conformity to building codes (Hansen/Vaa 2002). These three forms of 'illegality' often overlap. Moreover, living and income standards (informal) vary considerably, rendering these settlements extremely heterogeneous (Durand-Lasserve 1998a). The residents of informal settlements are generally involved in many activities, both formal and informal, and are not necessarily poor. In fact, informal settlements do not arise merely as a result of self-subsistence, but are increasingly part of a process of commercialization, a housing market that involves small- and large-scale owners. Some, because of poverty and the limits imposed on the local housing market, seek out modest dwellings in unauthorized settlements. Others are able to secure a dwelling and a source of 'informal' income by renting out part of their space. But there is no shortage of residential areas with medium to high living standards that have been developed in an informal manner, thanks in part to the push for privatization caused by structural adjustment policies.

The main difference between informal and formal settlements is the absence, in the former, of institutionally recognized infrastructure for the provision of water, sanitary services, roads, garbage collection, education, etc.

In addition to settlement modalities defined as informal, the cities of sub-Saharan Africa are also characterized by informal economic activities, undertaken outside the normative frame of reference (Castells/Portes 1989: 12, in Hansen/Vaa 2002: 10). Beginning with the decrease of so-called formal jobs in the 1970s due to the contraction of the public sector and limited industrialization, informality became a crucial aspect of the African city. The '*informal sector*'[5] grew considerably during the 1980s, and in the 1990s it absorbed at least half the work force of African cities (Simone 2004: 25). According to UN Habitat (2009), informal employment activities represent 60 % of work in urban areas, and that percentage is even higher if only women are considered.

While informal activities were initially valued for their level of independence and flexibility, a growing number of informal workers are now employed by formally organized economic operators (Simone 2004: 25).

3.2.2.2 The Question of Informality in Urban Studies

Many studies have addressed the question of informality from a variety of perspectives, and some of these have underlined the close interdependence and unclear distinction between formal and informal (Lee-Smith/Stren 1991: 27; Roy 2005,

[5]The 'informal sector' formula was introduced into urban studies by the *International Labour Organization* (ILO) in the 1970s. According to the ILO, the formal sector consists of large-scale enterprises that have a certain amount of capital at their disposal, while the informal sector is composed of autonomous workers and primarily provides livelihoods to the new inhabitants of the city. One of the most widely used classifications derives from the ILO report on Kenya, where the informal sector is characterized by easy access to local resources, family owned businesses, small-scale activities, unregulated competencies acquired outside the formal education system, and competing markets (ILO 1972). The ILO studies have highlighted the poverty of these activities, and their importance to the informal sector in terms of job and income creation (Sethuraman 1981, in Hansen/Vaa 2002: 10).

2009). The research of Roy (2005) on urban informality highlights how apparently contradictory approaches, which view informality as a plague on the expanding city of the Global South (Hall/Pfeiffer 2000) or as a resource that is able to offer creative responses to the state's incapacity to meet the needs of the most vulnerable people (De Soto 2000), are in fact similar insofar as they share a conception of the informal as a phenomenon that is separate from the formal. The majority of approaches that assimilate poverty and informality describe informal phenomena as a collective and local subsistence economy, isolated from global capitalism, although there are cases in which the informal economy is closely linked with the global one. Moreover, they ascribe the responsibility for poverty directly to the poor, who maintain a system of self-help among themselves and obstruct the role of the state because they view it as unnecessary. Roy/AlSayyad (2004) maintain, on the other hand, that the informal sector is not distinct from others; rather, a series of operations connect different economies and spaces to others.

In the vast literature on low- and medium-income countries, which offers a different, deeper and analytic interpretation of informality, three main contributions can be identified.

First, informality is found within the State's sphere of application and not outside it. It is often the power of the State that determines what is informal and what is not (Portes et al. 1989). In many cases, the State often operates in informal modes, benefitting from a territorial flexibility that purely formal mechanisms of accumulation and legitimation cannot offer.[6] For example, rapid peri-urbanization is an informal process that often violates state plans and norms, but it is also informally sanctioned by the State (Roy 2003). This means that informality is not an unregulated domain; rather, it is structured through various forms of extra-legal, social and dialectic regulation.

Second, informality is much more than an economic sector; it is a modality of production of space (Roy/AlSayyad 2004). "Informality produces an uneven geography of spatial value thereby facilitating the urban logic of creative destruction. The differential value attached to what is 'formal' and what is 'informal' creates the patchwork of valorized and devalorized spaces that is in turn the frontier of primitive accumulation and gentrification. In other words, informality is a fully capitalized domain of property and is often a highly effective 'spatial fix' in the production of value and profits" (Roy 2009: 826).

Third, informality is internally differentiated. The splintering of urbanism does not occur in the fissure between formality and informality, but rather in the sphere of informal production of space. With the consolidation of neoliberalism, a 'privatization of informality' also took place. While informality was once predominantly situated on public land and practiced in public space, today it is a fundamental mechanism of all privatized urban areas and is integrated into the market, as occurs with informal land parcelling that creates the 'peri-urbanization' of many cities (AlSayyad/Roy 2004: 4). These forms of informality are no more

[6]To reword a term coined by Brenner (2004), these too are '*state spaces*' (Roy 2009: 826).

legal than unauthorized (squatter) settlements and shantytowns, but they are an expression of 'class power', and can thus command infrastructure, services and legitimacy in a way that renders them substantially different from the slums (Roy 2009: 826).

Such questions are obviously at the centre of the debate on African cities, and on the Global South in general, where informality is the principle modality of the production of space, but they are also relevant for all cities, because they draw attention to several key characteristics of urban processes: "the extralegal territoriality and flexibility of the state; modes of social and discursive regulation; and the production of differentiated spatial value. In this sense, informality is not a pre-capitalist relic or an icon of 'backward' economies. Rather, it is a capitalist mode of production, par excellence" (Roy 2009: 826).

From this perspective, informality is not understood as an object of regulation by State, but as a product of the State itself. Roy (2005: 149) uses the concept of 'state of exception' and 'sovereignty' to explain this point, referring to the work of Carl Schmitt and Giorgio Agamben, who sees sovereignty as the power to determine the state of exception. The paradox of sovereignty consists in being simultaneously within and beyond juridical order (legitimate and non-legal); in this sense, informality can be considered an expression of such sovereignty. It follows, then, that legalization of informal property, of the spatial management practices and informal modalities of accessing resources are not simply technical or bureaucratic problems, but become complex political problems.

Although urban informality is seen as a problem of urban development, often to be addressed through political responses such as requalification and concession of land use rights (regularization) (Gulyani/Bassett 2007; Peyne 2002), the identification of what challenges and paradoxes urban informality creates for planning can help to change the perspective, to better understand the relations between formal and informal, and to initiate specific evaluations of the dominant political responses that have often generated negative impacts. The informal as opposed to the formal could be rethought, according to several authors, as a strategic instrument for planners, who may see in it opportunities to mitigate certain conditions of vulnerability of the most sensitive people (Roy 2005: 147).

3.2.3 Urban Development, Environmental Deterioration and Environmental Transformation

There is a double interaction between urban development and environmental changes. On the one hand, urban development has been guided over time by the characteristics and changes of the natural environment, and this has been particularly evident in Africa. On the other hand, increasingly rapid growth of the city generates notable impacts on the environment and natural resources, upon which, as we have seen, the lives of the people who live in hybrid rural–urban settlements of sub-Saharan Africa directly depend.

The development of sub-Saharan cities is often associated with environmental degradation and pollution. The modality with which cities develop has modified the natural environment and has influenced the relationship that people and communities have with it, both in terms of access and management of natural resources and in terms of the cultural and symbolic value attributed to them.

Several studies have revealed how new cities during the colonial period were born in an environment that was already deteriorating (Freund 2007: 79–80). The African neighbourhoods that developed in proximity to the centre of the city were poor and unplanned, characterized by small, very simple (poor) rooms intentionally built for individual workers. Shanties, mud and garbage seem to be the elements of African cities since colonial times, both in the ghettos created in the centre of cities and in the outlying zones that grew up around it (Freund 2007).

Geographical and environmental factors, particularly climate change, have always influenced the urbanization process in Africa more than in other contexts. The process of urbanization has therefore also been the result of the flight from difficult environmental conditions rather than a migration towards cities viewed as attractive for their economic and social opportunities (Annez et al. 2010).

The emergence of problematic questions such as Climate Change necessitates a rethinking of the urbanization process not only as a threat but also as an opportunity.[7]

If, as climate scenarios suggest, the temperature continues to rise and precipitation increases in the tropical zone of Africa (IPCC 2007, 2013), there will be increasingly frequent and intense climate changes that threaten to make the people who straddle urban and rural economies and environments even more vulnerable, as well as those in the high-density settlements situated in zones of high environmental risk (humid, depressed, or coastal zones, etc.). Moreover, urban development and variations in socio-economic conditions in peri-urban areas alter exposure and sensitivity to environmental change, affecting the probability that a given environmental change will occur and orienting the impacts thereof in a positive or negative way.

Faced with obvious environmental changes, the inhabitants of peri-urban areas are diversifying their livelihoods, in some cases reducing their dependence on natural resources through increased dependence on urban occupations and services. This diversification provokes changes in social relations, values and livelihood priorities that political decision-makers and planners should consider when envisioning the future of African cities.

Several key questions for planning have emerged from the analysis of the peri-urban interface. First, the criticality and environmental impacts generated by urban development are linked, in addition to social justice, to the sustainability of peri-urban ecosystems vis-à-vis pressures placed by the urban region on renewable

[7]Changes in rainfall patterns, for example, may represent an advantage in some cases, because it allows for an increase in the number of possible cultivation cycles over the course of a year, and facilitates agriculture in urban areas with temporary modalities.

and non-renewable resources and on environmental services. In this sense, the environmental sustainability of urban and adjacent rural areas conditions, and is conditioned by, flows of goods, merchandise, capital, natural resources, people and the pollution that they produce.

A second question concerns the environmental conditions of the interface as a living and working environment of a large number of people. As has already been mentioned, the peri-urban interface, with its rapid and heterogeneous changes to social composition, constitutes the habitat of small farmers conditioned by dynamic processes of land use and market changes, and of low-income groups who are varied between each other, are occupied in livelihood activities linked to the urban centre, and live in informal settlements in the 'periphery' of the city. These conditions render the people who live in the peri-urban interface vulnerable to the impacts and negative externalities of the nearby urban and rural systems. Such people are often subject to the worst effects of both worlds, since they are often located in areas that are subject to flooding and environmental risk, and where there is a lack of access to water and basic sanitary services.

The need to consider these two questions in an interconnected way is evinced by the principles and objectives outlined in the United Nations Conference on Environment and Development (UNCED) Agenda 21, signed by 100 governments in 1992 in Rio de Janeiro. Four years later, the Habitat Agenda—the Istanbul Declaration—confirmed and expanded Agenda 21, calling for effective action in order to provide adequate housing to everyone and 'sustainable human settlement' in an urbanizing world. Agenda 21 and the Habitat Agenda are two milestones in a significant change in perspective that occurred during the 1990s, because they recognize environmental problems as an integral part of social and economic development processes. This change has also shed new light on the role of urbanization and development processes and their vast impact beyond the city limits.

3.3 Dar es Salaam: Formation and Development of Peri-urban Areas

Dar es Salaam was founded in 1862 in a coastal region that was previously occupied by small clan settlements of the Zaramo tribe, which lived principally from fishing and agriculture (Leslie 1963, in Šliužas 2004: 7). Born as a port city serving the Sultanate of Zanzibar's commerce between the East and the West, it came under the control of the German East Africa Company in 1887, and then in 1891 of the German government itself (Lonsdale 1992), which established the capital of what, at the time, was the state of Tanganyika. The British government, which took over from the Germans in 1916, confirmed the city's role as the capital, which it would retain even after independence in 1961, until the capital was transferred to Dodoma in 1974. Both the construction of the Empire of Zanzibar and the German and British empires were determinative in defining the structure of the city.

In its 150 years, Dar has been characterized by initially inconsistent demographic growth, which allowed it to attain the title of 'metropolis'.

Today, Dar is the most populous city (4,364,541 according to the 2012 census[8]) in Tanzania and the third-fastest city in Africa as regards demographic growth (4–7 %). In 2002, it contained 33.7 % of the entire urban population of Tanzania, much of which resides in the municipality of Kinondoni (UN Habitat 2010: 12). The expansion of settlements that is associated with such growth has occurred, as in the majority of sub-Saharan African cities, in an 'informal' manner (Fig. 3.2). Dar es Salaam is also an example of the socio-economic and cultural changes connected to the process of economic and political liberalization begun at the end of the 1980s, which gave a new impetus to the urban economy after the economic crisis. The distribution of profits from this economic recovery were concentrated, as often happens, in the hands of the few, and the city underwent a radical change from a commercial perspective, thanks to the introduction of western products and the diffusion of instruments of communication and local media (Brennan et al. 2007).

Although great attention is being paid to environmental questions, studies on the city of Dar es Salaam have concentrated mainly on issues such as health and sanitation, and only in small part on the interaction between natural and constructed environments, or the hybrid characteristics of urban areas.

The relationship between lifestyle in urban areas, environmental management and the use of resources has received relatively little attention, which has prevented both recognition of the dynamics that characterize peri-urban areas and determine their present configuration, and identification of the relationship between these and other aspects, such as the land use regime, development policies, upgrading programmes, etc.

3.3.1 *The Evolution of Peri-urban Space in the Colonial and Post-independence Era*

3.3.1.1 The Period of European Colonization, 1887–1961

Between 1887 and 1916, Dar es Salaam and much of the surrounding area were administered according to the German regulatory system. In 1891, the German government decided to move the capital of German East Africa, or Tanganyika, from Bagamoyo, 60 km to the north, to Dar es Salaam. A series of new buildings was therefore constructed north of the port and around the train station, opened in 1905, which were intended to house new administrative functions and to give a growth stimulus to the city.

[8]Source: National Bureau of Statistics, TZ, at: http://www.nbs.go.tz/.

Fig. 3.2 Aerial view of Dar es Salaam central business district and surroundings. *Source* Ricci (2014)

During this period, the majority of the settlements for labourers were located in spontaneous settlements in the margins of the planned city (Sutton 1970). Although the Germans recognized the need to improve inhabitants' living conditions, they concentrated their attention on the city centre, initially preparing a development plan for only the Kariakoo area, and subsequently expanding it to include the entire nucleus of the urban centre.

In fact, through zoning and city planning norms, the Germans implemented a form of racial segregation of the urban space, a direct expression of colonial power. This process consisted of diverse and specific instances of encapsulation of the peri-urban areas (of that period) within the city, evidence of a complex interaction between the centre and the surrounding areas that it is impossible to classify (Brennan et al. 2007: 2). Some areas were, in fact, encapsulated within the city as rich suburbs (Upanga, Oyster Bay), while others became densely built-up African neighbourhoods (Kariakoo, Gerezani, Ilala), hemp plantations (Msasani peninsula) or areas for households of farmers (Buguruni, Kijitonyama, Chan'gombe). Some peri-urban neighbourhoods of the period were at the centre of a system of relations that involved a broader territory, while others remained completely isolated for the entire twenty-first century and did not reflect the national growth level (Brennan et al. 2007: 2).

During British domination (1916–1961), the population grew consistently. In 1939, it was noted that workers' living conditions were very modest, there was a high level of unemployment (25 %), poor education, and many people lived in temporary and overcrowded houses (Iliffe 1970). The population grew rapidly after the Second World War, and from 1947 on the formation of slums and shantytowns was considered out of control (Leslie 1963: 22). In 1956, the population of Dar reached 130,000 people. The colonial administration responded to this growth with the adoption of a Master Plan, but such measures were insufficient to improve the government of urban expansion or to satisfy the need for new housing (see Sect. 4.3). In fact, although a variety of new neighbourhoods were constructed, the settlements of the African population (the most numerous) remained mostly spontaneous and without infrastructure.

3.3.1.2 Growth After Independence

The population continued to grow after independence in 1961, aggravating the problem of providing adequate housing and services. Although the new national local government recognized the importance of managing the growth of the city and responding to the population's need for housing and services, the strategies adopted during this period were unsuccessful. According to several analyses, in 1992 35 % of the built-up area of Dar es Salaam consisted of informal settlements where 60 % of the population resided. More recently, other authors have maintained that the population of informal settlements had reached 70 % (Kombe 2005: 115; Lupala 2002b: 28; Kombe/Kreibich 2000: 40; Kyessi 2002), while others still suggest that 80 % of settlements were located in unplanned areas (Kironde 2006: 46).

As in many sub-Saharan African cities, the dimension of informal development has been and remains an important issue for the management of the city, together with the question of rapid population growth.

The economic conditions under which the rapid growth of the city occurred in the post-colonial period severely limited the opportunities of many households, contributing to an increase in urban poverty (Šliužas 2004: 73). Various attempts were made to confront such questions through political and regulatory instruments, whose effects on peri-urban areas are briefly discussed in the following paragraph.

3.3.1.3 Spatial Organization and the Colonial Legacy

The vestiges of colonial regulations are apparent in the organization of space. Although Dar es Salaam underwent radical transformations following independence in terms of physical and demographic expansion, post-colonial development of the city occurred largely according to the trends established during the colonial era, and spatial segregation continues to be evident (Brennan et al. 2007).

Residential areas, intended for Europeans (*Uzinguni*) and Indians (*Uhindini*) during the 1950s and composed of high-quality single dwellings, are still predominantly inhabited by communities that are not native people. The presence of Indians and well-to-do Africans after independence could be considered a change, but the continuities are even clearer: the racial segregation of the colonial period has, since 1961, simply been replaced by segregation according to income.

Continuity with respect to the colonial period is also quite evident in the subdivision of commercial space. In fact, the area in which Indians and Asians live today continues to be the residential and commercial centre of the city.[9]

Another important aspect of colonial segregation was the city–country distinction: while rural areas were left to African men and especially women, the cities were perceived as spaces for non-natives (Mbilinyi 1985, in Brennan et al. 2007: 5). Although residency rights were also recognized for African groups, limitations on their potential to reside in urban areas persisted, even after independence.

Finally, even the division of resources between the zones occupied by the Indian, African and European communities is considerably unequal, with the areas for African residents receiving lower investment in terms of infrastructure and services.

Segregation and discrimination, in addition to determining spatial organization, have also played an important role in the growth of political conscience and action (Mbilinyi 1985). In fact, the interaction between Indians and Africans led to the birth of anti-colonial, and subsequently post-colonial nationalism.

[9]The Kariakoo neighbourhood, despite having been transformed in recent years by the introduction of tall buildings, is still today a place of exchange with and between Africans. The city centre (Mijini), on the other hand, is where the financial and commercial institutions owned by non-Africans and government officials are located (Brennan et al. 2007: 4).

3.3.2 *From Structural Adjustments to the New Face of the 'City'*

In order to understand the evolution of peri-urban areas in Dar es Salaam, one must understand how the reforms of Tanzania's socialist past and those determined by the transition to neo-liberal economy have combined with the inheritance of the colonial period (see Chap. 3). The peri-urban areas in Dar es Salaam have, in fact, undergone a series of increasingly significant modifications since 1961, during the socialist period of the first president, Nyerere, the subsequent economic crisis, and the adoption of Structural Adjustment Programmes promoted by the International Monetary Fund in the 1980s and 1990s.

Such programmes have aggravated inequalities in Tanzania, particularly in Dar es Salaam, where situations of extreme wealth and poverty in the same context are often seen. Those same programmes have facilitated the development of a burgeoning financial market and have introduced elements that fuelled urban growth in areas that are increasingly distant from the city centre.

These changes have influenced spatial structures, migration models and the use of land and urban and peri-urban agricultural areas (Nelson 2007: 32).

3.3.2.1 Growth of the Informal Peri-urban Economy

The transition from a socialist to a market economy began in Tanzania in 1981 with the National Economic Survival Programme, but the changes to economic policy were limited while Nyerere was in office. It was with the presidencies of Ali Hassan Mwinyi (1985–1995) and Benjamin Mkapa (1996–2005) that significant structural change has come about with, for example, "the sale to the private sector of parastatal companies, the proactive encouragement of domestic and foreign capital, and the privatization and introduction of competition to the financial sector" (Briggs/Mwamfupe 2000: 801–802) The current President, Jakaya Kikwete, who came into power in 2006, has continued to emphasize the privatization of services and foreign direct investment, following the economic process initiated by the adoption of Structural Adjustment Programmes.

Recent economic reforms have led to two main types of change that have had an impact on the peri-urban areas of Dar es Salaam.

The first type of change was introduced by the Zanzibar Declaration in 1991, which marked a radical shift from the socialist focus on *Ujamaa*[10] towards recognition that "not all informal activities to supplement incomes were

---

[10]In Swahili, *Ujamaa* means 'extended family'. According to this principle, which was at the basis of Nyerere's socialist policies, a person becomes what they are through other people and the community.

undermining the political order and the economy, and that many activities involved the creation of new products and services were vital to the survival of urban dwellers" (Tripp 1997: 188).

The recognition of a dynamic and expanding informal sector allowed many people to find regular employment in informal economic activities in order to improve their own livelihood capacity. Prior to this legislative act, informal activities such as peri-urban agriculture were actually prohibited, or based on the rigid principles of the *Ujamaa*.

The second change was the liberalization of the transport[11] system in Dar es Salaam, which favoured residential development in areas that were previously inaccessible due to their distance from the city centre and main thoroughfares[12] (Briggs/Mwamfupe 2000).

3.3.2.2 Changes in Territorial Structure and Peri-urban Migration

In light of such transformations, migration towards peri-urban areas follows two distinct tendencies that influence the definition of the peri-urban environment.

The first is related to the movement, from urban to peri-urban, of medium- and high-income households. Their transfer to peri-urban areas has been favoured not only by improved accessibility due to the liberalization of the transportation system, but also to the existence of housing subsidies provided to government employees, and to the availability of low-cost land, which has led to a veritable construction boom in certain peri-urban areas (Briggs/Mwamfupe 2000).

For these reasons, peri-urban areas have become increasingly residential, in conflict with agricultural use, and maintain social and economic connections to the central areas of the city. For medium- and high-income people from urban areas, investments in construction have constituted a means of overcoming economic difficulty.

The second tendency involves the young people of rural areas who migrate to Dar. For them, the possibility of accessing a house or other dwelling, work, and other resources is based on the social networks present in peri-urban areas (Kombe 2005). In addition to this important social factor, the cost of living in peri-urban areas is much lower.

Certainly, the combination of the effects of the economic crisis and Structural Adjustments led to increased commercialization of the land in peri-urban areas over the course of the 1990s (Briggs/Mwamfupe 1999: 269) and transformed such areas

[11]A majority of the state-owned buses have been substituted with small minivans (*daladala*), which are privately managed.

[12]Morogoro Road, Bagamoyo Road and Pugu Road are the three managers of transportation towards the north and the west of the city, where the majority of urban growth is occurring. Growth has been limited to the south of the Dar es Salaam port in the Temeke municipality, which is not well connected to the rest of the city.

from “a zone of survival” to “a zone for investment” (Mbiba/Huchzermeyer 2002: 120). The consequence of this process has been the progressive exclusion of the poorest urban groups, and the aggravation of conflicts over access to land, which is also attributable to the regulatory transition from customary to formally institutionalized forms of land tenure.[13]

3.4 An Interpretation of Dar es Salaam’s Peri-urban Areas

The peri-urban areas of Dar es Salaam were, and continue to be a subject of interest for many researchers, even if the majority of studies are informed by ‘upgrading/rehabilitation’ goals, and therefore focus on poverty reduction or the resolution of hygienic and sanitary problems through the provision of infrastructure, neglecting rural–urban spatial patterns and relations.

Many researchers studying the growth of peri-urban areas have characterized them as informal development zones of the city, analyzing the dynamics related to the informal economy (Kombe 1995, 1999; Kombe/Kreibich 2000; Šliužas 2004; Ebebe 2011), urban agriculture (Sawio 1993, 1994, 1997; Kyessi 1997; Jacobi et al. 1999), and migratory phenomena (Lupala 2002b; Kombe 2003). Others highlight the crucial question of transition from an informal to a formal and more typically urban land tenure regime, which in these areas is manifest in all its complexity (Kironde 1995, 2003; Kyessi et al. 2009). Others still have studied settlement morphology (Lupala 2002a), the hygienic and sanitary system and the infrastructure of these areas, and have addressed the complex issues of planning and environmental management (Kombe 2005; Halla/Majani 1999).

One aspect that continues to be almost completely unexplored is the relationships between urban and rural that traverse the peri-urban area and determine its physical, social and economic equilibrium. The identification of elements that would be useful to the investigation of that relationship was one of the objectives of the questionnaire administered during the field research of the present study (see Sect. 3.4.2).

3.4.1 From ‘A Zone of Survival’ to ‘A Zone for Investment’: Peri-urban Agriculture and the Urban Market

Due to the economic decline of the formal sector, peri-urban agriculture has, since the 1980s, been a crucial activity for the livelihood strategies of three quarters of the

[13]See *Land Act 1999* and subsequent legislation.

inhabitants of Dar (Lupala 2002a; 194). Estimates suggest that more than 25 % of the food consumed in Dare es Salaam was produced in peri-urban areas (Sawio 1993: 1).

In the peri-urban areas of today, agriculture is the main source of income for many households, while for others it contributes to a reduction in their food expenses, and allows them to sell their surplus production in the urban market (Lupala 2002a: 194–197).

Urban expansion has not, therefore, undermined the central economic role that agricultural activities play for the majority of people, despite the fact that it has been prohibited by several legislative provisions and must compete with other uses that attribute a greater value to the land (Fig. 3.3). In fact, the conversion of agricultural land for urban use has offered the owners of large and medium plots the possibility of selling or renting a portion of their land, but only a very small percentage of them has decided to definitively abandon agricultural activities; most prefer to keep part of their land for their own use as a source of income/production (Lupala 2002a). So-called 'part-time' agricultural activities, born of the necessity to compensate for the reduced buying power of urban salaries, are widespread in peri-urban areas. This allows peri-urban households to access non-agricultural employment thanks to their proximity to the urban centre, while many poor residents of urban areas are able to find employment in the agricultural sector of neighbouring peri-urban areas (Fig. 3.1).

Fig. 3.3 Urban agriculture along Morogoro road, Ilala Municipality. *Source* Ricci (2009)

Urban and peri-urban agriculture in Dar es Salaam, in addition to playing an important role in the economy and individual livelihood strategies, also adds value to unused or deteriorated land and improves certain areas of the city. It represents an alternative method of increasing land value in areas subject to environmental risk that cannot be used for housing; it prevents settlement and protects the land from parasites, the accumulation of refuse and vandalism. It can contribute to the restoration and improvement of soil conditions, increasing the use and rental value; it can increase the production of food, and therefore contributes to improved food security. Moreover, in the majority of cases it is an informal, spontaneous and creative activity, and is based on traditional knowledge and local techniques (Howorth et al. 1998: 24).

Agricultural practice in peri-urban areas coexists and often conflicts with the purchase and sale of land by middlemen, an activity that is at high risk of speculation[14] and is interwoven in the complex relation between formal land use title (leasehold for 33, 66 and 99 years, and customary tenure) and informal access to land for production or housing.[15] The purchase and rental value of plots, and the related speculative pressures, depend heavily on the location of plots. Those that are close or adjacent to road systems are more expensive because they facilitate movement and because they are located in strategic areas for potential small businesses and stores. The economic activities that can be undertaken in the peri-urban areas situated near main roads are the most disparate, and include stores for the sale of artisanal products or activities, temporary fruit and vegetable stalls, and others.

Given the rapid population growth and expansion of the city, many authors see the total formalization of land access in peri-urban areas as difficult (Lupala 2002b). The majority of the population of Dar is under 45, while the elderly constitute a much smaller portion. This implies an elevated demand for land that will necessarily be satisfied by the informal market, which is particularly active in peri-urban areas.

The challenge is therefore to understand the mechanisms and functioning of the informal sector, through investigation of land parcelling processes, and identification of key actors and local modalities of organization within which they operate. Thus, one must ask how the formal system can operate in harmony with the informal one.

These observations suggest, first, the increasing importance of the peri-urban economy, which results from rapid urbanization in conditions of poverty and is a

[14]Various strategies are being implemented in order to contain speculation. One strategy includes the planting of drought resistant plants; another entails the construction of small, temporary houses, sometimes consisting of a single room; a final strategy consists of reaching agreements with neighbours in order to protect plots and defend them against the possibility of occupation (Lupala 2002a).

[15]In Tanzania, all land belongs to the President, and formal access is regulated by the granting of leasehold titles (33, 66 or 99 years), and customary titles in rural areas.

fundamental support for environmental management in peri-urban areas. Second, they suggest that informal environmental management is closely linked to two main questions, one related to the real estate market and security of ownership, and the other related to land use regulation.

3.4.2 *Rural–Urban Interdependence and the Relationship with Natural Resources*

The field research carried out in Dar es Salaam reflects the fact, already mentioned by several authors (Allen 2006; Simon et al. 2004; Tacoli 1998a: 3), that the conception of the rural population as predominantly employed in agricultural production, while the urban population is employed is industry, is misleading. It also reveals the presence of flows of people, goods, and refuse between urban and rural areas, and the connected flows of money and information (Tacoli 1998b: 154–160) that are at the basis of people's livelihood strategies. The interactions between so-called rural and urban activities and spaces have been analyzed according to a variety of scales and perspectives, highlighting how rural activities often take place in areas that are defined as urban, and vice versa (Tacoli 1998b: 158).

It is in the peri-urban areas surrounding the more densely built-up urban area that these flows and the interaction between urban and rural are most concentrated and intense. Some of these relationships are represented by people who regularly travel from the area surrounding the city to the centre, by the flows of construction and other materials, by the flow of and plants for the treatment and storage of solid and liquid waste produced in the more densely constructed urban area, by the location of certain services and by building and land speculation (Tacoli 1998b: 158).

These interactions are often seen as the result of the city's growing influence on the land tenure regimes, land use, economic activities and labour market in the surrounding non-urbanized areas. This influence has significant impacts and on the livelihoods of the people who live in such areas, and determine their means and capacity to react to environmental, social and economic changes.

In order to investigate the interactions between peri-urban and central-urban areas, a section of the questionnaire was dedicated to exploring several contradictions that emerge from the assumptions most broadly used when defining policies and plans for sub-Saharan African cities—which we have defined as 'dominant interpretive approaches'—and the observations in the literature on the evolution and dynamics that characterize the peri-urban areas of Dar es Salaam.

3.4.2.1 Diversity

The first assumption analyzed is represented by the belief that peri-urban areas house mostly poor, uneducated people, without formal employment, who cannot afford to live in more expensive urban areas. On the contrary, the field research and questionnaires administered to the inhabitants of the peri-urban areas of Dar indicate great social and economic heterogeneity.

Almost all people work in agriculture, but this is never their only livelihood activity, and in some cases it is not even their main activity, but is combined with other types of work, in institutions, in commercial or business activities, in animal husbandry and fishing. This also reflects the weakness of a second assumption, namely that all people who live in peri-urban areas are engaged in agriculture as their primary activity.

Other evidence of diversity is provided by the analysis of education level. More than half the people interviewed had a primary education (Standard (STD) VII[16]), while the rest have a secondary education (FORM IV), and some a degree.

The questionnaire allowed for specific analysis of the differences between those who work exclusively or predominantly in agriculture, and those who are more involved in other sectors of activity.

As regards the 'length of settlement' variable, agriculture is the primary activity for the interviewees who have been settled for the longest period (more than 10 years), while those settled for less time also or only engage in other sectors (commerce and small business or employment in local institutions).

There is no correlation, on the other hand, between plot size and main livelihood activity. There are cases in which people occupied predominantly in non-agricultural activities had large tracts of land to cultivate that were used to varying degrees. By the same token, it also occurs that individuals whose livelihood is primarily based on agriculture have relatively limited arable land (less than one hectare). In general, the dimensions of cultivated land corresponds less to number of hours dedicated to agriculture, and more to the distance from the city centre. More specifically, the households with larger plots were located in less densely built-up areas (in the Madale and Bunju A wards) that were further from the urban centre and the coast, in other words, in areas where there was less speculative pressure and land parcelling occurred in an unplanned way.

Nor does the dimension of a plot correspond to the income level of households,[17] which consisted an average of 6 people. It should nevertheless be mentioned that not all the interviewees agreed to respond to this questions, and in some cases they claimed to have such a variable income that they could not accurately estimate their monthly earnings. The majority of those who responded to this question claimed to have a monthly income of between 20 and 150 US dollars.

Income level is also unrelated to the main type of activity (agricultural or non-agricultural), but it bears mentioning that the amount of money that a family had coming in (their income) was generally not indicative of their level of

[16]The levels of education are divided as follows: Primary Education (STD I–VII); ordinary level of secondary education (Form I to Form IV); advanced level of secondary education (Form V to Form VI); and university education (Bachelor or Masters degree).

[17]Three monthly family income brackets have been identified: from 20 to 150 dollars; from 150 to 300 dollars; and from 450 to 600 USD dollars.

well-being, except perhaps in the case of households with incomes over 450 dollars per month. Households' 'economic' condition depended on activities that entail the exchange of goods or services. In areas where there was no water supply system, for example, water could be exchanged for food o other goods. In other cases, the sharing of means of transportation or work can constitute another element of well-being that is not linked to income production. Most of these mechanisms are based on the presence of a system of social relations and networks that are difficult to outline but fundamental for people's lives.

3.4.2.2 Transition

The third assumption being questioned is the belief that peri-urban areas are the first landing point of migratory movements from the country towards the city (Mattingly 2009). The data analysis indicates that in the three Kinondoni district wards involved in the study, only a third of the households, and the overwhelming majority were from the Kinondoni district. It would therefore appear that transfers occur mainly within single districts and over short distances.

The reasons that the interviewees had settled in peri-urban areas are mostly linked to the desire or the need for a greater availability of arable land and space for animal husbandry, or for other activities linked to natural resources. A small number of people moved for work-related reasons that were not of their choosing (for example, transfers requested by their employer), to search for opportunities for self-employment, or to undertake a specific kind of business. About a third of the interviewees moved, on the other hand, for family reasons (to follow or re-join their family, or to start their own), as a result of governmental decisions beyond their control, or for other reasons linked to work opportunities. A few transfers were prompted by the government's concession of land titles for uncultivated areas, intended to favour the 'reclamation' and utilization of resources in remote areas.[18]

In fact, peri-urban areas attract a large number of people who often come from other peri-urban areas that are becoming urbanized, or from nearby or more distant rural areas, in search of a place that allows them to begin or continue agricultural activities, and to be closer to urban opportunities (markets, services, etc.) without having to respect the regulations imposed in the urban regime (tax payment, limitation of agricultural practices, limited open space, etc.).

Moreover, nearly all the interviewees have been settled for a relatively short period of time: less than ten years. This was mainly the case for people who moved to find a place to practice agriculture. This may indicate that the choice to work in agriculture is a contingent one, and does not exclude a future switch to urban employment. However, interviewees did not appear to be planning such a change,

[18]For example, several households that live in the Makongo area have been settled there for more than 20 years, benefitting from government incentives to occupy areas that are still in their natural state and to promote environmental upkeep and management in those areas.

and did not express the desire to move to urban areas where they could undertake urban activities and abandon agriculture (see *What Urban Myth?*).

Finally, it bears mentioning that in the mid 1990s (see the 1997 topographical map) many of the areas in which the interviews were carried out were almost completely rural. Consequently, there are very few households that have been settled there for more than 30 years. Nevertheless, the high level of dynamism in these areas is also the result of the public policies, discussed in the previous paragraphs, have often forced or pushed people to move many times in a short period, from urban areas to peri-urban areas or between different peri-urban areas, as a result of speculative pressures and the expansion of the central-urban nucleus.

3.4.2.3 Interdependence

The fourth assumption that has been explored holds that the inhabitants of peri-urban areas are completely dependent on the city for services and work. However, the present research has found that this kind of total 'parasitic' dependence, caused by the fact that peri-urban areas do not offer any kind of service or work opportunity, does not exist. Rather, there is a reciprocal interdependence between urban and peri-urban areas. While for certain services (e.g. filing requests for land titles) peri-urban inhabitants are forced to travel to the urban centre, for others (e.g. ordinary sanitation services, clinics) they can rely on local structures, formal or informal. The same is true for employment opportunities.

The field research and the literature on peri-urban areas of Dar es Salaam suggest that the desire to live outside the city, in areas that still maintain a rural character, is combined with stable interests and relations with the urban centre. The movements from peri-urban areas to the city, and vice versa, are frequent, and imply the existence of activities, flows and social relations based on urban–rural links, goods and strategies (Kombe 2003). Two thirds of interviewees claimed that they travel to the city centre on a weekly basis. This frequency increases in the wards that are closer to the central nucleus, such as Kawe, reflecting a closer relationship with the city. There are two possible explanations for this: the peri-urban areas that are located closer to the city are more densely built-up, and the lower availability of land for agriculture necessitates the search for livelihood activities in the urban area; secondly, proximity to the city facilitates movement between the two areas.[19] Only a small percentage of the households interviewed travels to the city centre on a daily basis, and a similarly low percentage travels rarely (twice monthly or less often). This suggests that there is a close interdependence with the urban area, especially for certain services linked to institutions, all of which are concentrated in the city centre.

---

[19]It bears mentioning that Kawe is located in proximity to two main city roads: Bagamoyo Road and Old Bagamoyo Road, which are fundamental for both internal transportation and for extra-urban and interregional movement.

Certainly these movements are conditioned by the availability of modes of transportation. Only some of the people interviewed possessed their own vehicle; nevertheless, movements are possible thanks to the presences of *daldala*, which make regular stops in almost every part of the region. This transportation modality is used by nearly all the households interviewed, and is integrated in the most remote areas with motorcycles (*pikpiki*) or three-wheelers (*bajaji*) that serve as local taxis.

People's reasons for going to the centre are predominantly linked to work and access to services. Many households, who carry out agricultural, animal husbandry or fishing as their principle livelihood activity often travel to the centre in order to sell their products.

3.4.2.4 What Urban Myth?

The fifth assumption under review concerns the belief that the inhabitants of peri-urban areas have 'urban dreams' that they want to live in a 'modern' city, and the fact that they live in peri-urban areas is imposed by economic and social conditions beyond their control.

The households interviewed were asked what their future expectations were, if they wanted to move to a completely urbanized area with full services, in which they could undertake activities other than agriculture and animal husbandry. The answer was negative in all cases, and motivated by the desire to live in contexts with open and green spaces where they could continue their agricultural and animal husbandry practices, a factor that had also determined their choice of the area in which they currently reside. No one expressed the desire to move to the city centre or another more urbanized area with more infrastructure; the few people that did express the desire to move are young, and claimed to want to go to areas where it would be possible to continue or increase their agricultural activities. This group includes young people residing in the Kawe ward of Makongo, close to the city centre, and people from Bunju, an area more distant from the central nucleus. In the first case, the need to move may be linked to the pressures of the urbanization process and the requalification projects planned in the study area (which will probably reduce, if not eliminate entirely, the possibility of carrying out 'rural' activities). In the second case, it is linked to the desire to engage in agriculture in other areas as a result of the reduced soil productivity in their current location, or to return to their region of origin.

3.4.2.5 Peri-urban Priorities

It is clear that the assumptions discussed above can orient the development of peri-urban areas in one direction rather than another, and this can introduce or remove elements that interfere with people's livelihood strategies. These questionnaire responses strongly indicate that access to land as an asset to cultivate and as space for a peri-urban lifestyle is a priority, while the need for infrastructure for the provision of water and other types of urban services are secondary concerns.

This is because there is a system of (informal) organization of people that allows for the substitution of services offered by infrastructure (waste collection or water supply) through people's relationships and labour, which render peri-urban areas a space in which to live, albeit in a dynamic way, and not one from which to flee.

This does not mean that these areas function well by themselves, or that their relationship with the urban area is balanced and sustainable. Rather, the goal of this study is to highlight the fact that an awareness of the quality and quantity of relationships (in terms of importance for people's lives) can be a crucial point of reference when formulating interventions for improving these areas. While the priority must be recognition of the fact that the relations and actions (platforms) on which people's livelihood strategies are based cannot always be substituted with infrastructure (which, moreover, can upset the delicate balance between the urban and rural environments as well as access to resources), it is equally important to prevent informal modalities of management of space and alternative provision of services from damaging the environment and human health, or causing social injustice.

The informal organization for collecting certain types of solid refuse can substitute the work carried out by public waste collection services, but monitoring of the final destination of the waste collected is essential for avoiding environmental disasters. The informal distribution of water is crucial in areas that are mostly without water supply networks or that lack continuous and widespread water provisioning, but it is equally important that illegal access, maintenance of the infrastructure present and the potential speculation linked to the sale of this resource are monitored. Moreover, the 'platforms of action' referred to above contain and reproduce power relations that could generate mechanisms of oppression and exploitation as a result of different economic and cultural positions; in this case, again, awareness and intervention by institutions can be very useful in avoiding such mechanisms.

3.5 Reconsidering 'Modern' Approaches to Urban Development: From 'Neo-Colonialism' to Sustainability

The new approaches adopted by urban policies in the post-colonial era have masqueraded as strategies for poverty reduction and the promotion of sustainable development, however they have tacitly favoured the persistence of a kind of 'colonial regime' through land speculation and the exploitation of resources (e.g. tourism, infrastructure building, etc.), particularly in peri-urban areas.

In the international debate on the management of urban growth in sub-Saharan Africa, and elsewhere, there is a growing consensus as regards the need to develop planning approaches oriented to promoting sustainability, inclusion and visibility, a clear move away from the notion of planning that was concentrated on spatial order and control (Todes 2011: 115). New approaches must be understood in terms of the contemporary urban reality and planning theories that rethink the very nature of planning and its relationship with power and institutions, and which view cities as complex and dynamic places comprised of multiple interests and spaces.

From this perspective, the goal is to identify the challenges that contemporary sub-Saharan cities pose to planning, and determine where to focus research into new planning processes that will reduce vulnerability and environmental change.

The above reflections raise several emblematic questions for further consideration. The first of these is that of the link between infrastructure, people's livelihood strategies and spatial planning. This issue is linked to the ambiguity of the idea of the 'compact city' that is generally opposed to peri-urban areas and defines them as unsustainable models of soil sealing. Finally, there is the transversal question of sustainability, more specifically sustainable environmental management (carrying capacity, equilibrium of natural resource cycles as well as economic sustainability and social justice). While this last issue will be discussed in Chap. 3, in terms of policies, planning and their contribution to reducing vulnerability of environmental changes, the first two will be discussed in the second half of Chap. 4, where both infrastructure (informal) and the non-compact and hybrid structure of peri-urban areas is evaluated and discussed in terms of their contribution, as positive as it is negative, to peri-urban area residents' adaptive capacity to environmental change.

3.5.1 New and Old Challenges in Planning

Redirecting attention to the processes that characterize urbanization in Africa means shedding light on the conflicts related to access to, and control and use of resources. Such conflicts, most of which are linked to land, play a crucial role in processes of the production of space, and as such they cannot be ignored by planners and city administrators.

We have seen that the people who live in peri-urban areas have a close link with natural resources, and that their livelihood strategies adapt according to the environmental changes that occur. Environmental management therefore represents a fundamental point of intervention for institutions seeking to improve the environment, and thus people's lives, in peri-urban areas, and to mitigate the conflicts that may result therefrom. As we have seen, peri-urban areas are characterized by informal economies and settlements that manage the environment in an autonomous way. The informal production of space and the state of exception that this implies are produced by the State, and this occurs at various levels, in both gated communities as an informal expression of the settlement of the medium and upper classes, and in shantytowns. This implies that planning is directly involved in these phenomena, and that addressing informality means confronting the ways in which planning produces the unplanned and the unplannable (Roy 2005: 156). To move from a functional land use planning towards distributive justice, to rethink the object of development and to substitute models of good practice with realistic criticism are not only political considerations with which to approach the informal, they also indicate that the informal is an important source of knowledge for urban planning (Roy 2005: 156).

If one considers predominantly informal peri-urban development not as something to oppose and control, but something that planning should take into consideration in the terms suggested by Roy, reducing vulnerability means, first and foremost, addressing the question of social justice. In other words, attention must be paid to the fact that spontaneous urban development processes can also be guided by power imbalances (economic, social, etc.) and therefore can aggravate situations of exploitation or self-exploitation. Second, reducing vulnerability means recognizing the fact that informal (and spontaneous) production and management of space can have negative effects on the environment and on natural resources. This compromises the systems of resource management and access on which the lives of certain groups of people are based, especially in areas with hybrid rural–urban characters, and who depend directly on a relationship with natural resources. Finally, it means overcoming the idea that there is some kind of 'natural self-regulation' of the informal, and requires specific evaluation of situations in terms of the impacts they have on people's lives.

(a) Expansion of hydro and sewer networks:
Dar es Salaam Water Supply and Sanitation Project (DWSSP);
(b) New industrial settlements:
Kigamboni oil refinery project.

Although the 1979 Plan initiated the process of questioning several positivist and structuralist planning principles, interventions and provisions that effectively consider informal development of the city as an 'exception' and a negative alteration of urban development are evident in subsequent development programmes and plans, which sought to intervene through regularization and formalization[20] as the only solutions for reducing people's vulnerability to environmental, economic and social conditions. The policies and interventions that have been implemented come into conflict with a dynamic and rapidly growing city, in which so-called informal settlements and environmental management practices, as well as the presence of hybrid–rural–urban activities and forms, are a widespread and diverse reality that characterizes the modalities of production of space in the city. This therefore requires specific consideration.

[20]A formalization program for economic activities and the use and occupation of land is also present in Tanzania (MKURABITA), which is oriented to defining the areas to formalize, informing and educating citizens and city leaders in order to promote the process, defining regularization schemes that must be approved by governing authorities, and surveying and registering lots for the concession of occupation certificates. This process is not favoured by those who continue to buy, sell, and use land in an informal way (see Chap. 5).

3.6 Which Environmental Transformations: Global Changes and Local Effects in Sub-Saharan Cities

We have seen how the cities of sub-Saharan African are characterized by rapid growth that is unaccompanied by the development of ‘urban’ infrastructure and dwellings, by serious socio-economic disparities, by limited governability and environmental degradation. These conditions increase sensitivity to environmental transformations, and climate change can worsen environmental, social and economic situations that have already reached a critical level.

The main impacts on sub-Saharan African cities include those deriving from the occurrence of extreme events that are striking the tropical belt of the African continent in an increasingly intense manner (El Nino, storms, severe drought, flooding, etc.), and the environmental stress connected to the availability of water, which is a serious strain on people dependent on natural resources and on the food production system, and therefore food security. In addition, heat waves, air pollution and flooding can heavily condition the health of the urban environment, leading to serious impacts on people’s health, which are also linked to the spreading of vectors of disease according to climate variations. Finally, there are numerous impacts on rural–urban connections at the regional (between rural and urban areas) and local level (between urban areas and the peri-urban fringe). Access to resources and the possibility of diversifying income sources is often based on these connections, which can therefore be thrown into crisis as a result of environmental stress with consequent conflicts over resource use and migratory phenomena on various scales.

Dar es Salaam is subject to a variety of environmental changes linked to the combination of its geographical and environmental characteristics and the modalities with which the city has and is being developed.

3.6.1 Environmental Transformations and Climate Change in Dar es Salaam

In Dar es Salaam, the problems caused by climate variability (especially flooding of the coastal zone during the rainy seasons) have been at the centre of the urban agenda for some time. Extreme events such as the 2004 Tsunami or El Nino (1992–1993 and 1997–1998) (Shemsanga et al. 2010, cited in URT DEO 2011) have produced a series of damages to coastal infrastructure. In addition to continued stress linked to climate change, including the modification of rainfall patterns and the acidification of the ocean with its destructive effects along the barrier reef, several phenomena of environmental degradation that are already underway as a result of the rapid urban development process (water pollution due to the lack of purification of urban sewage, intense fishing and agricultural activities conducted with improper modalities, multiplication of water extraction wells) are expected to worsen (coastal erosion and salinization of groundwater) (Dodman et al. 2009).

The essential data on the main climate variations that are relevant for the city are listed as follows:

- Rainfall variability: there has been a decrease from 1200 mm annually in the 1960s to 1000 mm in 2010. Furthermore, increased unpredictability of rainfall has been noted, with temporal shifts in the rain peaks and notable impacts on agriculture and people's livelihoods;
- Temperature increase: the minimum temperature level was below 20 °C between 1979 and 1986, while it rose to in 21 °C 1989 and between 20 and 21 °C in 2000, and since 2000 it has continued to increase (TMA 2011). The average temperature increased by 0.2 °C between 1979 and 2009, while the average minimum temperature between 1979 and 2008 increased by 2 °C.
- Rising sea levels: the global rise revealed by the IPCC (2007) was about 3.1 mm annually during 1993–2003, while that rise decreases to 1.8 mm annually if one considers the longer period of 1961–2003, reflecting an acceleration in recent decades. This is particularly significant for Dar es Salaam and coastal ecosystems (mangroves, coastal wetlands, cliffs, vegetation, tides, etc.) and may, according to some authors, continuously modify the coastline, destroying coastal infrastructure and pushing inhabitants to migrate. Such transformations are already evident in some areas of Dar es Salaam, such as the Kunduchi ward on the northern coast.

There considerable uncertainty regarding climate variations, but the sea level rise is particularly controversial. Indeed, measurements of sea level show a pattern of decrease in Dar es Salaam (Kebede/Nicholls 2010).

3.6.2 Planned Adaptation and the Role of Local Institutions in Dar es Salaam

The data analysis, as well as several other studies (Dodman et al. 2011; UTR DEO draft 2011), suggests that residents, especially in the peri-urban areas of Dar es Salaam, are aware of the environmental changes underway. In order to face some of those changes and to facilitate environmental protection (e.g. tree planting), residents are also implementing strategies, which are useful to varying degrees in mitigating the impacts of environmental change (Mbonile/Kivelia 2008).

Several similar strategies, though not explicitly connected to adaptation objectives, are being adopted at the institutional level, including guaranteeing the provision of infrastructure and services for the poorest populations. Nevertheless, an 'adaptation deficit' still persists in the city (Dodman et al. 2011).

The government of Tanzania had already adopted the National Environmental Action Plan in 1994 (UTR 1994), laying the foundations for the subsequent national environmental policy. Tanzania took its first step towards addressing the issue of climate change with the ratification of the UNFCCC in 1996, which was concretized with the drafting of the National Adaptation Programme of Action (NAPA) in 2007. The NAPA recognizes the need to incorporate the impacts of future climate change into public policy.

Many of the adaptation initiatives concern rural areas; nevertheless, in recent years several studies have concentrated on Dar es Salaam (UN Habitat 2011; Dodman et al. 2011). During the COP 15 in Copenhagen 2009, the Mayor's Task Force on climate change chose Dar es Salaam as one of the cities where pilot studies would be conducted. The Task Force will concentrate on three aspects: understanding the links between urban poverty and climate change, identifying best practices for reducing urban poverty and vulnerability to climate change and promoting investment programmes in order to support best practices. Other programmes related to climate change that are currently underway include Climate Adaptation through Participatory Research and Local Action, in the Temeke municipality, which is expected to expand into two other municipalities; and the Dar es Salaam Resilience Action Plan (DRAP), which should lead to the development of additional actions related to climate change and the integration of the results of various on-going studies that are part of the new Master Plan.

Furthermore, the Dar es Salaam city administration has a Disaster Management Unit that coordinates responses to disasters, including those related to climate change, such as fighting cholera epidemics.

Following the publication of the NAPA, a series of studies were initiated in order to assess the city's vulnerability to climate change, and they are still in progress.

An analysis of the plans and the relatively limited strategies related to climate change indicate that it has been difficult to implement them and to find resources. Interviews with employees of local institutions (Kinondoni district and Kunduchi, Bunju and Kawe wards) have revealed, on the one hand, the existence of a large number of plans for environmental conservation and protection, and on the other, that most city planning is still concentrated at the ministerial level, and long-term planning is almost entirely absent. In fact, the interventions in which those employees are involved are generally very specific and linked to responses for resolving contingent problems.

Some plans at the national level that pursue environmental conservation objectives can nevertheless contribute to adaptation to climate change. This is the case as regards the General Plan for the Conservation of Natural Resources, drafted by the Ministry of Natural Resources and Tourism, to be implemented by municipalities in order to preserve the mangroves, control fishing and mariculture practices, and to prevent illegal fishing, unauthorized sand pits, and unjustified tree cutting. Although these are not explicitly objectives of adaptation to climate change, many of the planned measures can also reduce the vulnerability of people and ecosystems to environmental change. However, it should be noted that the environmental conservation measures have never been situated in relation to urban development measures, which are principally linked to the development of infrastructure. As a result, institutional behaviour is often contradictory.

3.7 A New Environmental Question for an Old Planning Problem

We have seen that the environmental, social and economic effects of environmental change are considerable. The planning policies and strategies underway leave many questions unanswered, which draws attention to several criticalities that are not new to urban studies.

The issue of environment has been at the centre of the contradictions between contemporary urban strategies, as is reflected in Marvin and Hodson's criticisms of securitization approaches (2009). In fact, it would appear that the theories of urban bias and anti-urban bias still constitute the pivotal points around which such strategies revolve, especially for peri-urban areas that continue to be seen as incomplete pieces of the city, or parts of the countryside to reclaim and restore.

Even the issue of informal development of the city is brought to the centre of climate change: that which is informal 'doesn't exist', or exists as a problem to resolve, reclaim and regularize. On the other hand, the informal is also a world of resources and creativity that is capable of self-organization in any situation, and can survive even when institutional contributions are limited.

The planning system in Tanzania has been developed (at least in theory) around a tradition of 'comprehensive planning', first introduced by British colonizers, and later reinforced by the Master Plans developed after independence (MacGregor 1995). In practice, this approach is often substituted and/or supplemented with planning for single interventions, guided by an interpretation that pays little attention to processes and directives in a complicated and fragmented combination of regulations. Given the inefficacy of plans and projects, planners have had to ask themselves why development processes did not follow their long-term forecasts, trapped in the dilemma of having to either 'accept' reality or impose norms.

The attempts to respond to these problems in the disciplinary sphere have reflected several changes, including a growing tendency to place less emphasis on the planning of limitations and control, and a view of planners' input as only one of many inputs necessary in the development process. This has led to new forms of non-technical knowledge and the involvement of community members interested in the definition of a common vision (Healey 1997, 2010). Planning is increasingly understood and practiced as an iterative, participatory and flexible process, despite the fact that in many cases it still proceeds as a process that keeps the analysis of urban and regional transformations separate from the governance processes through which decisions are made.

As we have seen, the environmental conceptualization of the peri-urban interface and the vast interactions between rural and urban that give rise to specific transformation processes of that area have a variety of implications as regards the formulation of planning interventions. The complexity of the phenomena and subjects that characterize the peri-urban interface poses numerous questions for planning, questions of scale, instruments, competency and others.

The development of a new perspective implies a rethinking of intervention policies and strategies: a 'non-urban' interpretation of the peri-urban interface could modify how certain political questions are formulated (the informal, for example) and how certain environmental problems are represented.

How one can arrive at a planning approach that responds to the specificity of the peri-urban areas remains an open question that is widely debated. One contribution that may allow that debate to move forward may come from research into a new interpretation of space that could be the basis for identifying sustainable processes that would need to be preserved in order to develop new ones, which will be essential to improving the living conditions of peri-urban and other inhabitants.

While 'securitization' and urban development represent an indiscriminate solution for any type of urban and peri-urban space, some authors assert that careful analysis of the local community, consideration for participatory models in the public sphere and an understanding of how transformations can be triggered and/or managed is necessary in order to orient planning to the benefit of the most vulnerable groups. A focus on strengthening the organizations of peri-urban residents is also essential in order to render their hopes and needs 'visible' to the various institutions that intervene in this context. Nevertheless, in order to prevent an 'isolationist' approach from impeding the emergence of independent coalitions of diverse social groups, examination of the 'political capacity', both individual and collective, of peri-urban groups and its effects on the management of resources, is necessary (Allen 2006).

Chapter 4
Environmental Management and Urbanization: Dar es Salaam as an Illustrative Case

Abstract This chapter concentrates on approaches to environmental management and urban planning that have become skewed over time in order to resolve the African 'environmental crisis' in the name of sustainable development and/or poverty reduction. Such approaches have been brought to the fore by the global debate on Climate Change. The extent of the environmental transformations currently underway in Dar es Salaam and sub-Saharan Africa more generally are discussed, as well as the manner in which the two global strategies of mitigation and adaptation to Climate Change orient urban development policy and planning at the local and global level. The adaptation strategy emerges as crucial to planning processes in African cities, prompting a reconsideration of the impact of strategies that emphasize 'securitization' of the city as opposed to acceleration of the rural–urban transition in order to reduce social vulnerability. Such strategies raise questions that are not new to the planning debate, and draw attention to the role that people must play therein.

Keywords Climate change · Adaptation · Spatial planning · Securization · Governance · Local institutions · Brown/green agenda

4.1 Approaches to Urban Environmental Planning and Management in Sub-Saharan Africa

In the previous chapter, we have seen how the questions of sustainability and environmental management in sub-Saharan African cities and peri-urban areas have been included in the debate on urban development since long before the Rio Conference in 1992, with the disappearance of the concept of 'over-urbanization' and criticisms of the urban bias theory and the green revolution. The failure of the latter leads one to recognize that the living conditions of the population, in both urban and rural spheres, depend on proper environmental management.

L. Ricci, *Reinterpreting Sub-Saharan Cities through the Concept of Adaptive Capacity*, SpringerBriefs in Environment, Security, Development and Peace 26, DOI 10.1007/978-3-319-27126-2_4

The paradigm of sustainable development, as well as neoliberalism and processes of privatization and institutional decentralization and devolution, has guided the development agenda in contemporary African cities. Sustainable development has taken on a variety of meanings in international documents (McGranahan/Satterthwaite 2002; Buckingham-Hatfield/Percy 1999, cited in Myers 2005) that generally refer to the balance between economic growth and environmental impacts. This study does not propose to address the various interpretations of the concept of sustainable development; rather, it is concentrated on how that concept is translated into policies and interventions to transform the city.

Mayer warns us that in sustainability there is the risk of inadequate consideration for the political dimension, which is intentionally omitted in order to conceal problems related to ideological interests and conflicts or power relations. In this sense, economists' attempts to construct sustainable development models based on the absence of negative change in the stock of natural resources and environmental quality, come up against the need to understand who defines what a negative change is (Myers 2005). So, what appears to be a method for addressing the negative effects of industrialization and urbanization and improving social, economic and environmental conditions, becomes an instrument that serves the market, top-down planning, scientific and technological solutions based on infrastructure projects intended to resolve environmental problems configuring as a new form of neoliberalism[1] (Myers 2005).

In contrast, for the majority of African urban studies, sustainable development underlines the necessity of sustainable living conditions on the local level (Lerise 2000; Rakodi/Lloyd-Jones 2002; Myers 2005) and refers to the Sustainable Livelihoods Approach. This perspective also orients the reasoning of those who maintain that in order to understand the application of dominant approaches to sustainable development, such as the UN Sustainable City Programme, one must consider cities' environmental governance policy networks and their environmental consequences for the livelihoods of ordinary city residents (Myers 2005; Allen 2006). The following paragraphs analyze the policies and programmes, driven by the concept of sustainability, that have oriented urban development in Africa.

[1]At the World Summit on Sustainable Development in Johannesburg, 2002, the U.S. Secretary of State at the time, Colin Powell, defined sustainable development as "a new approach to global development, designed to unleash the entrepreneurial power of the poor" with "good governance, sound institutions, economic reform, transparency in your system, the end of corruption, responsible leadership, responsible political activity, and […] decision-making based on sound science" (Powell 2002: 6–7, cited in Myers 2005).

4.1.1 From the Emergence of the Environmental Debate to Sustainable Cities: 'Brown' and 'Green' Agendas

As we have seen, the debate over the relationship between environmental issues and urban development in peri-urban areas of African cities is oriented by two typologies of question: one is related to the environmental conditions that affect people's livelihoods and quality of life (especially the poorest); the other concerns the environmental sustainability of pressures imposed by urbanization processes on renewable and non-renewable resources and environmental services (Allen 2006). The former is linked to reflections on peri-urban areas as places of rapid and heterogeneous change, inhabited by people whose livelihoods depend on the opportunities offered by a living environment with mixed urban and rural characteristics, which renders them simultaneously vulnerable and rich in terms of resources. In the second case, peri-urban areas are at the centre of flows between urban and rural areas of products (goods, merchandise), capital, natural resources, people and pollution, which condition and are conditioned by environmental sustainability.

The two approaches translate into policies and interventions that correspond to two perspectives on urban planning and environmental management, defined 'green' and 'brown' agendas. The 'green' agenda concentrates on long-term environmental problems that derive from the impacts of development, such as rainforest reduction, global warming and the loss of biodiversity, and is presented as a way to reduce the impact of urbanization and industrial production. These challenges are more frequent in the North, and often have long-term effects (McGranahan/Satterthwaite 2000). The 'brown' agenda highlights the need to focus on specific problems associated with the deterioration of local environmental conditions, with consideration for questions related to the degradation of households' living environments, unhealthy living conditions, waste, contamination of water sources, the lack of drainage systems, etc. These questions have immediate consequences for health and the environment, especially for low-income people (Leitmann 1999) and are therefore primarily concerned with the countries of the South.

Although, condition of degradation of households' living environments exist also in both in developed and developing countries, these two agendas are often considered separately, perhaps not from a theoretical perspective, but in practice and at the level of political debate, where there is a tendency to focus either on local environmental problems that have an immediate and obvious impact on people's health and quality of life, or on problems related to the sustainability of natural resource management (Allen 2006). One of the main reasons for this dichotomy lies in the difference between the epistemological and ethical orientations adopted by the two agendas: 'defence of nature and ecology' in the case of the green agenda, 'people, rights and places' in the case of the brown agenda (Allen/You 2002).

The need to consider both perspectives in an interconnected way has been underscored by Agenda 21, drafted at the 1992 United Nations Conference on

Environment and Development (UNCED) in Rio de Janeiro, and confirmed 4 years later by the Istanbul Habitat Agenda (Habitat II) in 1996, which expanded upon Agenda 21 and called for effective action to provide adequate housing for all and sustainable human settlement in an urbanizing world. Agenda 21 and Habitat II are two milestones in the significant change of perspective, which occurred over the course of the 1990s that recognized environmental issues as an integral part of social and economic development processes. This change shed new light on the role of urban development processes in development and on their impacts within and beyond the city limits. Building on the Habitat II, the next bi-decennial cycle lead to the Habitat III United Nations Conference, to be held in Equator in 2016, should give a boost to the global commitment to sustainable urbanization through the implementation of a 'New Urban Agenda'.

Recognition of the fact that brown and green agendas are, by definition, interconnected led to a renewed debate on the environmental problems and opportunities that arise from the rural–urban interaction. Although this is not a new area of interest, it is giving rise to new interpretations of the African city and the traditional concepts of 'urban' and 'rural' that reveal the inadequacy of modalities with which the processes of environmental change that characterize the peri-urban interface are defined and addressed (Allen 2006, 2009).

Moreover, the perspectives of Agenda 21 and the Habitat Agenda have reflected the link between sustainability and governance. Sustainable development is seen as the result of a process that involves common people in their daily lives, rather than the exclusive domain of governments and experts. As a result, the need to include the people who are usually marginalized and excluded from decision-making processes is emphasized, placing the question of participation at the centre of the debate over sustainable development.

Moreover, the forthcoming Habitat III and initiatives, such as the Sustainable Urban Development Network (SUD-NET),[2] are expanding the debate on the different dimension of sustainable urban development. Particularly, the SUD-NET, as network of global partners, focuses on promoting interdisciplinary approaches to sustainable urban development identifying the following priority themes: governance; environmental planning and management, urban planning; urban economy; and education, training and research. The SUD-Net followed the termination in 2007 of the Sustainable City Programme (SCP) and Local Agenda 21 programmes.[3]

[2]SUD-NET is an initiative supported by UN-HABITAT to promote sustainable urban development. It works at the local level to strengthen the capacity of national governments, local authorities and communities. SUD-Net is currently active in the Cities and Climate Change Initiative (CCCI) and Habitat Partner Universities (HPU) http://mirror.unhabitat.org/content.asp?typeid=19&catid=570&cid=5990.

[3]The Sustainable City Programme is a joint UN-Habitat/UNEP facility, established in the early 1990s to build capacities in urban environmental planning and management. Local Agenda 21 (LA21) aimed to support local authorities in achieving more sustainable development by implementing an Environmental Planning and Management process.

4.1.1.1 Green and Brown Agendas: A False Dichotomy?

Some critics (Shiva 1993; Atkinson 1994; McGranahan/Satterthwaite 2000; McCarney et al. 1995) have addressed the conflict between the two approaches (green and brown), asserting that the global agenda on environmental issues has created a situation in which certain problems identified by the Global North (and often created by the North) lead to the development of theories and innovations that are specific to the North, but oddly are also assumed to be transferable to the Global South.[4] Although it has been recognized that the environmental problems of the South are different (specific), when it comes to environmental planning theories or strategies, what is applied remains relatively unspecific or falls within one of the two dominant streams of planning.

If, on the one hand, the differences between the two agendas can generate conflicting approaches to environmental planning, on the other, it is argued that the dichotomy is irrelevant, since in many cities addressing brown agenda items also involves green agenda items. In other words, the problems linked to issues of health and environment (e.g. inadequate housing and services) and those related to industrialization (e.g. uncontrolled factory and transportation emissions) can be viewed as two aspects of the brown agenda (Leitmann 1999). McGranahan/Satterthwaite (2000: 74) maintain that it is important to avoid creating this false dichotomy and to direct attention towards greater equity in both agendas at a broader level.

Equity is a principle recognized by both agendas in their approach to sustainable development. In fact, one, placing the emphasis on intra-generational equity, recognizes that all inhabitants of urban areas have the right to a healthy and safe living and working environment and to the infrastructure and services related thereto; the other, assuming instead an inter-generational perspective, is more concerned with the possibility that urban development will fail to consider the finite nature of resources and cause ecological systems to deteriorate, compromising the capacity of future generations to meet their own needs (McGranahan/Satterthwaite 2000).

Although both agendas converge in terms of emphasizing the global environmental impact of unequal processes of consumption and production in various parts of the world, social and environmental justice require a notion of fairness that is not limited to addressing unequal distribution of goods and services, or in environmental terms, the unfair distribution of environmental goods. They must also consider the institutional structures and social relations that produce and reproduce such inequality of distribution.

[4]Some critics (Hajer 1996; Lafferty 2001; Myllyla/Kuvajab 2005, cited in Allen 2010) have defined ecological modernization, irrespective of issues of environmental injustice and inequality, as an adaptation effort by capitalism, intended to mitigate environmental impacts through modern means of production, in which the emphasis on technological change and management pursues efficiency within a market system. From this perspective, the problems caused and identified by the North should be resolved everywhere, even in the South, through the export of innovations produced in the North.

Some authors relate issues of distribution equity to political and economic strategies at the global level. Recent industrial restructuring processes of a neo-liberal bent have accelerated the competition over environmental resources and have promoted the maximization of short-term profits through mechanisms that are socially and environmentally unsustainable. This has led not only to negative impacts on local livelihood strategies and deterioration of environmental conditions, but also to a weakening of the standards regarding natural resources due to the competition between national and foreign capitals, and between labourers and the state (Allen 2001). The neo-liberal policies of the Structural Adjustment Programmes are inserted into this framework (see Chap. 2), with a notable impact on the relationship between production and nature, and have changed the institutional landscape in most of the cities in the Global South.

Global commerce has cause cities to become less dependent on their hinterlands in terms of livelihoods, and at the same time the refuse produced in urban areas is exported to distant regions. This often means that the origin of food and energy and the destination of waste is invisible to urban inhabitants, and creates a dependence that could generate geopolitical and ecological instability and unsustainability (Allen et al. 2007). The limits imposed by the expansion of the urban ecological footprint do not become evident until they translate into effects at the local level, such as rising food and energy costs, frequent flooding or an increase in disease related to pollution of the environment. At first, this appears to be particularly true in the North, since the cities of the South seem to still maintain a strong dependence on their hinterland. Nevertheless, a more detailed analysis reveals that expansion of the ecological footprint is also linked to income level, which means the question of sustainability and the link between brown and green agendas cannot be addressed without examining the rural–urban relationships and inequalities produced by contemporary urbanization processes in the global North and South (for example, the relationship between hyper-consumerists and sub-consumerists;[5] Allen 2010).

4.2 Global Climate Change as a Driver in the Debate on Environmental Transformations and Settlement Processes

The environmental debate begun in Rio in 1992 already contained the theme of climate change, but the question of environmental transformations connected thereto has emerged in urban studies only in the last decade, as the deadlines established in the Kyoto Protocol drew closer. The long debate on the relationship between urbanization, development and sustainability was therefore reoriented, with climate change as the central dimension of sustainability, which reignited

[5]These assumptions have led to the development of approaches that are centred on themes such as the city-region and the urban bioregion.

discussion of environmental justice and compensation between countries of the North and South.

Climate change is recognized at the international level (IPCC 2007) as one of the most important and complicated challenges for twenty-first century society, and scientific evidence about global warming and impacts have been broadly addressed. The Fifth Assessment Report of the Intergovernmental Panel on Climate Change (IPCC) states that warming of the climate system is unequivocal and each of the last three decades has been successively warmer at the Earth's surface than any preceding decade since 1850, the atmosphere and ocean have warmed, the amounts of snow and ice have diminished and sea level has risen (IPCC 2014a). Many aspects and areas of human life are affected by such changes. Agriculture, water resources, biodiversity, industry, health and cities, and the impacts of climate change will not be uniform around the world, rather considerable differences are anticipated according to region (McCarthy et al. 2001). The poorest nations, which contributed the least to greenhouse gasses emissions, are the most vulnerable to climate change (Huq et al. 2003). For this reason, special funds have been created in order to help the poorest countries react to the causes and impacts of these changes.

The policies and strategies for addressing environmental change in the cities of low- and medium-income countries aim primarily to reduce people's vulnerability by orienting approaches to spatial planning, seeking to reduce the causes of climate change (mitigation) and, above all, reducing the impacts of environmental transformation that are already under way (adaptation).

The mainstreaming of adaptation into policy and planning processes assumes a central role in the struggle to reduce social and biophysical vulnerability in the city, but it also represents a critical aspect. If planners do not distance themselves from the approaches that have oriented urban planning up to now, which are based on the 'asymmetrical ignorance' that characterizes many urban studies, the mainstreaming of adaptation in current practices runs the risk of exacerbating the effects of environmental change rather than reducing them.

In fact, recourse to interventions for 'completion', securitization, sanitation and upgrading of the 'not yet cities' in sub-Saharan Africa as a solution to the challenges posed by climate change clashes with processes of settlement and spatial management that are temporary, hybrid, rapid and multidirectional and actually give form and body to contemporary space, especially in peri-urban areas. This clash is explanatory as regards the conflict between global ideas of the city, which are at the basis of policies and instruments that often increase social vulnerability to environmental change, and local processes of the production of space.

The re-emergence of the environmental question and the need to include adaptation objectives to environmental change in urban planning therefore seem to render even more evident the contradictions inherent in the imposition of western planning models and the 'marginality' to which such models relegate the capacity to act and environmental management of members of the community (e.g. active social networks in lieu of conventional services and infrastructure).

4.2.1 Climate Change: From Geneva to Lima, Toward Paris

In Geneva, 1979, the First World Climate Conference indicated climate change as an absolutely urgent problem at the global level, calling on governments to anticipate and pay attention to climate risks. At that time, the Global Climate Programme was started, guided by the World Meteorological Organization (WMO), the United Nations Environment Programme (UNEP) and the International Council for Scientific Unions (ICSU). A variety of intergovernmental conferences on climate change followed, and in 1988, in Toronto, 46 countries participated in the Conference on the Changing Atmosphere, from which the necessity to develop a convention framework in order to protect the atmosphere emerged. The WMO and UNEP thus founded the Intergovernmental Panel on Climate Change (IPCC) in order to evaluate the totality and the speed of changes, their impact and strategies for addressing them. The First Assessment Report of the Panel was published in 1990, with important repercussions for political decision-makers and public opinion. In the same year, the Second World Climate Conference was held, which established an Intergovernmental Negotiating Committee to draft a framework convention on climate change. At the Rio Conference in 1992, the United Nations Framework Convention on Climate Change (UNFCCC) was presented and ratified, which established an action framework aimed at stabilizing atmospheric concentrations of greenhouse gasses in order to avoid 'dangerous anthropogenic interferences' with the climate system. The convention was not originally legally binding on the nations that had ratified, since it did not impose mandatory limits on the emission of greenhouse gas. However, it did include updating provisions (referred to as 'protocols') that could impose mandatory limits on emissions, the most important of which is the Kyoto Protocol (COP[6] 3, 1997). Two parallel and complementary strategies were formulated to mitigate the impacts of environmental transformations linked to climate change (adaptation) and to address the causes of that change by reducing the greenhouse gas emissions that cause global warming (mitigation).

The most recent Conferences of the Parties have sought to orient funds and programmes for adaptation and mitigation not to the production of new ad hoc instruments (plans and projects) but to the integration of measures and strategies into development processes in low- and medium-income countries, and into the planning and management of cities (adaptation mainstreaming).

[6]The Conference of Parties (COP) is the highest decision-making authority of the UNFCCC, and all countries that adhere to the Framework Convention are members thereof. The COP is responsible for international efforts to address climate change, for examining the application of the Convention and the engagement of the Parties in light of Convention objectives, new scientific knowledge and acquired experience in implementation of climate policies. The COP meets only once per year. The work of the COP is supported by two subsidiary organs: the Subsidiary Body for Scientific and Technological Advice (SBSTA) and the Subsidiary Body for Implementation (SBI).

With the Kyoto Protocol, which came into effect in February, 2005, industrialized countries and those in transition to a market economy (192 parties) committed to reaching the objective of reducing emissions by 5.2 % with respect to the 1990 level during 2008–2012 (the first period), with specific objectives that varied from country to country (Annex I of the UNFCCC).

In 2005, the COP 11 and COP/MOP[7] 1, held in Montreal, established the *Ad Hoc Working Group for the Kyoto Protocol* (AWG-KP) to work on further commitments by the adhering parties to the Kyoto Protocol on the basis of Article 3.9. Countries agreed to consider long-term cooperation on the Convention through a series of four workshops known as the Convention Dialogue, which continued until COP 13.

In Bali, 2007, the COP 13 and the COP/MOP 3 led to the adoption of the *Bali Action Plan* (BAP), which created the *Ad Hoc Working Group on Long-term Cooperative Action* (AWG-LCA) with the goal of concentrating on the key elements of long-term cooperation identified during the Convention Dialogue: mitigation, adaptation, finance and technology transfer. The Bali conference was translated into an accord on biennial process, the Bali Roadmap, which established two sets of negotiations pursuant to the Convention and the Protocol, and established deadlines for the conclusion of those negotiations at the COP 15 and COP/MOP 5, held in Copenhagen in 2009.

Although the first milestone for the development of the adaptation regime under the UNFCCC is considered the *Least Developed Countries* (LDC) Work Programme established at COP 7 in Marrakesh (2001), followed by the Nairobi Work Programme (NWP)[8] and the *Cancun Adaptation Framework* (CAF),[9] in Bali, the question of adaptation emerged in a relevant manner with emphasis on the needs of the countries' most vulnerable to the negative effects of climate change. The adaptation issue reached a level of importance equal to that of mitigation, and the inextricable link between the two was recognized. The level of mitigation determines the level of temperature increase, and therefore the totality of climate change and the need to adapt thereto.

4.2.1.1 From Bali to Lima

In 2008 and 2009, two AWGs held four parallel negotiation sessions (Bangkok, Bonn, Accra and Poznan in 2008; Bonn, Bangkok, Barcelona and Copenhagen, in 2009). The goal of these meetings was to move forward with negotiations to reach an agreement on long-term cooperation in the COP 15 and COP/MOP 5 in Copenhagen. The conference in Copenhagen was marked by disputes over transparency of the decision-making process. An informal negotiating process between

---

[7]Conference of the Parties acting as a Meeting of the Parties for the Kyoto Protocol.

[8]Nairobi Work Programme (NWP) was established in Nairobi at COP 12 in 2006.

[9]Cancun Adaptation Framework (CAF) was developed in Cancun at COP 16 in 2010.

the most important economic actors had led to the political agreement known as the Copenhagen Accord. The Accord was also presented at the plenary session of the conference, and attendees were divided between those who supported it as a step forward for a better future, and those who held the negotiations hostage with complaints of a lack of transparency and democracy in the process. In the end, the COP 15 agreed to 'take note' of the Copenhagen Accord and established a process through which the Parties could express their support for the Accord. To date, more than 140 countries have expressed their support, and 80 countries have provided information on the objectives of reducing emissions and mitigation actions.

In Copenhagen, the decision was also made to extend the mandate to working groups on the Convention and the Protocol (AWG-LCA and AWG-KP), asking them to present their respective results at the subsequent COP 16 and COP/MOP 6, in Cancun, 2010. Many hoped that Cancun would be able to produce significant progress on several key questions, but the results were modest and legally non-binding. Negotiations on themes such as mitigation, adaptation, finance, technology, *reducing emissions from deforestation and forest degradation in developing countries* (REDD+), *monitoring, reporting and verification* (MRV), and international consultation and analysis (ICA), led to the drafting of the 'Cancun Agreements', which established several new institutions and processes, including the Cancun Adaptation Framework and the Adaptation Committee, and about which there are contrasting opinions and some opposition. The majority of participants recognized that it represented a relatively small step forward in the struggle against climate change. In addition to the Cancun Accords, 20 decisions on other issues were adopted that range from capacity building to administrative, institutional and financial issues.

After four more negotiating sessions, held in 2011 to prepare documents and reach compromises on the unresolved questions, many of the issues that arose in Cancun remained unresolved and awaited political input, leading to not binging Conference in Durban (COP 17). The Durban outcomes cover the establishment of a second commitment period under the Kyoto Protocol, a decision on long-term cooperative action under the UNFCCC and agreement on the operationalization of the *Green Climate Fund* (GCF) created in Cancun. Parties also launched the new Ad Hoc Working Group on the *Durban Platform for Enhanced Action* (ADP) with a mandate "to develop a protocol, another legal instrument or an agreed outcome with legal force under the Convention applicable to all Parties" (IISD 2012: 2). The negotiations of the 'Durban Platform', a set of decisions that lay the groundwork for adopting a legal agreement on climate change, is scheduled to be completed by 2015, with the outcome entering into effect from 2020 onwards.

After the Bonn Climate Change Conference and the Bangkok Climate Change Talks, held, respectively, in May and August 2012, the COP18 and the *Meeting of the Parties to the Kyoto Protocol* (CMP 8) was held in Doha. Negotiations in Doha focused on ensuring the implementation of agreements reached at previous conferences. The 'Doha Climate Gateway' package included amendments to the Kyoto Protocol that establish its second commitment period. The AWG-KP, launched at CMP 1 in 2005, terminated its work in Doha together with the AWG-LCA and negotiations under the Bali Action Plan. Key elements of the outcome also included agreement to consider loss and damage, such as institutional mechanisms to address loss and damage in developing countries that are particularly vulnerable to the adverse effects of climate change. While developing countries and observers "expressed disappointment with the lack of ambition in the outcomes of Annex I countries' mitigation and finance, most agreed that the conference had paved the way for a new phase, focusing on the implementation of the outcomes from negotiations under the AWG-KP and AWG-LCA, and advancing negotiations under the ADP" (IISD 2012: 1). At COP 19 and CMP 9, held in Warsaw in November 2013, the Adaptation Committee inter alia recalled that planning for adaptation should be based on nationally-identified priorities (Decision FCCC/SBI/2013/L.10/Add.1). The COP 19, under the Cancun Adaptation Framework, established the 'Warsaw International Mechanism for Loss and Damage', subject to review by COP 22, including structure, mandate and effectiveness to address loss and damage associated with extreme weather and slow onset events in developing countries that are particularly vulnerable to the adverse effects of climate change. The Warsaw international mechanism is tasked to enhance knowledge and understanding of comprehensive risk management approaches; strengthen dialogue, coordination, coherence and synergies among relevant stakeholders; and enhancing action and support, including finance, technology and capacity-building (Decision FCCC/CP/2013/L.15).

Although the loss and damage issue was one of the most contentious item of the conference, it was also unable to renovate developing countries' confidence that the UNFCCC process can meet their expectations.

However, the COP20 in Lima (December 2014) adopted the initial 2-year work plan of the Executive Committee of the Warsaw International Mechanism for Loss and Damage and finalized the organization and the governance of the Executive Committee. Moreover, negotiations in Lima focused on outcomes under the ADP necessary to reach an agreement at COP 21 in Paris in 2015, and on a draft decision for advancing the Durban Platform for Enhanced Action adopting the "Lima Call for Climate Action". Decisions adopted in Lima help operationalize the Warsaw International Mechanism for Loss and Damage. The Lima was able to lay the groundwork for Paris 2015, setting in motion the negotiations towards a 2015 agreement, the process for submitting and reviewing *intended nationally determined contributions* (INDCs), and enhancing pre-2020 ambition.

4.2.1.2 The Least Developed Countries: Policies, Objectives and Adaptation Instruments

While the first COP in 1995 addressed the issue of financing adaptation (decision 11/CP.1), it was only with the Marrakesh Accords of 2001 that adaptation received greater consideration (decision 5/CP.7). The international community recognized the vulnerability to climate change of *Least Developed Countries* (LDCs[10]) and their limited adaptive capacity, and as a result the UNFCCC differentiated roles and responsibilities between the Parties to a greater extent than was done regarding mitigation. The convention distinguishes between developed countries, developing countries and least developed countries, between vulnerable and particularly vulnerable, and between industrialized countries and those in economic transition. The convention also distinguishes between countries with different physical characteristics, highlighting the specificity of the needs of small islands, low-lying coastal areas and countries with fragile ecosystems. This differentiation is used to attribute different responsibilities to various groups.

Article 4 of the UNFCCC establishes that countries included in Annex II of Annex I (industrialized) in addition to undertaking mitigation measures, must also provide financing to help developing countries adapt to climate change. This took on particular importance following the Marrakesh Accords (COP 7, 2001), which established the Least Developed Countries Fund (LDCF) managed by the Global Environmental Facility[11] (GEF). The fund is intended to respond to the specific needs of LDCs, including the drafting and implementation of National Adaptation Programmes of Action (NAPAs),[12] whose purpose is to identify the priority needs of each of the Least Developed Countries according to specific guidelines provided by an expert ad hoc group (Least Developed Countries Expert Group—LEG).

After the publication of the IPCC Third Assessment Report in 2003 at the COP 9, the Subsidiary Body for Scientific and Technological Advice (SBSTA) was given the task of working on the scientific, technical and socio-economic aspects of vulnerability and adaptation to climate change (decision 10/CP.9). The following year, the Buenos Aires Programme of Work on Adaptation and Response Measures (decision 1/CP.10) was created, which established two complementary routes for adaptation: the development of a 5-year work plan on the scientific, technical and

[10]Article 4.9 of the UNFCCC recognizes the unique situations of Least Developed Countries (LDCs) and asserts "The Parties shall take full account of the specific needs and special situations of the Least Developed Countries in their actions with regard to funding and transfer of technology".

[11]Based on the provisions of the UNFCCC, the GEF manages three funds: the GEF Trust Fund, the *Least Developed Countries Fund* (LDCF) and the *Special Climate Change Fund* (SCCF). Further financing opportunities for adaptation projects in LDCs include the Adaptation Fund provided in the Kyoto Protocol, funds deriving from *Multilateral Environmental Agreements* (MEAs) and multilateral or bilateral funds from governments, organizations and national or international agencies. The GEF Trust Fund and its *Strategic Priority for Adaptation* (SPA) supports pilot projects and demonstrative activities that address adaptation while also generating environmental benefits. The purpose of the COP in terms of the GEF's support for adaptation identifies three phases. The first phase provides support to the national communication process, which includes assessment of vulnerability and adaptation. The second phase provides further assistance for other adaptive capacity building actions. The third phase concerns support for adaptation activities, including insurance.

[12]In order to address the most urgent issues, the UNFCCC established that LDCs should develop NAPAs that, recognizing current adaptive strategies at the local level, identify priority adaptation activities that can be implemented with the support of the LDC Fund.

socio-economic aspects of vulnerability and adaptation to climate change and improvement of information and methodologies; and the implementation of concrete adaptation actions, technology transfer and capacity building. At the COP 12, a preliminary list was compiled of activities to undertake during the 5-year period of the work plan, known as the Nairobi Work Programme on Impacts, Vulnerability and Adaptation to Climate Change, and the Adaptation Fund, established by the Kyoto Protocol, was modified. The IPCC Fourth Assessment Report and Working Group II's results as regards impacts, adaptation and vulnerability indicate that hundreds of millions of people will be exposed to an increase in stress on water supply, that millions of people will be exposed to flooding each year, and that access to food will be seriously compromised in many African countries, emphasizing that adaptation will be necessary but that many impacts can be avoided or delayed through mitigation.

We have seen how the Bali Acton Plan, adopted at the COP 13 (2007), identifies adaptation as one of the key components for strengthening future responses to climate change, and for allowing full, effective and lasing implementation of the convention through long-term cooperation up to 2012 and beyond. In the subsequent Copenhagen Accord (2009), institutional representatives (heads of state, heads of government, ministers and other delegation leaders) highlighted the need to establish a global adaptation programme. The signatories agreed that the development of actions and international cooperation on adaptation were urgent and that developed countries must provide adequate, predictable and sustainable financial resources, technology and capacity building in order to support the implementation of adaptation intervention in developing countries. Financing for adaptation would be assigned first to developing countries, such as LDCs, Small Island Developing States (SIDSs) and Africa.

In Cancun, 2010, a new Cancun Adaptation Framework was established and a Green Climate Fund was created in order to allow for better planning and implementation of adaptation projects in developing countries through an increase in financial and technical support, as well as the definition of a process for continuing to work on 'loss' and 'damage'. The implementation of adaptation in LDCs is therefore linked, on one hand, to financial and technological support and capacity building, and to the development of NAPAs and their implementation through local adaptation plans.

4.2.2 *Causes and Effects of Environmental Change: Two Strategies for the City*

According to the UNFCCC, climate change refers to the changes directly or indirectly attributable to anthropic activity that alter global atmospheric composition and influence the natural climate variability observed over comparable periods. On the other hand, Working Group II of the IPCC uses the term climate change in reference to any change in climate over time, resulting from both natural variability and anthropic activity (IPCC 2011: 21). From this perspective, the climate change currently being observed is the combined result of processes that generate changes in the concentration of *greenhouse gases* (GHGs)[13] and aerosol in the atmosphere, solar radiation and land surface characteristics (land and land cover). According to the IPCC Fifth Assessment Report (IPCC 2014c), there is evidence of that warming consistent with anthropogenic CC, has increased across Africa, where high temperatures are associated with increased mortality. This increase could trigger changes on an even larger scale over the next few decades, with elevated and non-linear impacts, possibly consistent across physical and biological systems and potentially irreversible. Although projected rainfall change over sub-Saharan Africa in the mid- and late twenty-first Century is uncertain and downscaled projections vary according to the topography and local environment, climate change is likely to be characterized by a reduction in precipitation in sub-Saharan Africa (IPCC 2014b). Moreover, African ecosystems are already affected by climate change, and future impacts are expected to be substantial. Water-stressed catchments with complex land uses, water infrastructure and stress on water availability will be amplified by climate change and socio-political and economic conditions. A major challenge is the coupled effect of climate change and fast urban growing.

4.2.2.1 The City as a Cause of Change

Emissions of carbon monoxide, the most important anthropic greenhouse gas, are predominantly due to the use of combustible fossil fuels, and secondarily to land use. They are therefore connected to the sectors of energy production, transport, industry, land use and changes in land use and forest management. Related to these sectors, and in particular to energy consumption, are population growth and per capita income, which between 1970 and 2004 rose by 69 and 77 %, respectively. The discrepancy in the contributions of low-income and high-income countries[14] to the production of greenhouse gas is therefore clear. The production of greenhouse gas is also highly differentiated at the national level, and in this respect the role of

[13]The work of the IPCC is divided between three Working Groups, a Task Force and a Task Group. The activities that each Working Group and the Task Force undertake are coordinated and administered by Technical Support Units. The IPCC Working Group II (WG II) assesses the vulnerability to climate change of socio-economic and natural systems, the negative and positive consequences of climate change and adaptation options. It also considers the correlation between vulnerability, adaptation and sustainable development. The assessment of information is conducted according to sector (water resources, ecosystems, food and forests, coastal systems, industry, and human health) and regions (Africa, Asia, Australia and New Zealand, Europe, Latin America, North America, polar regions, and small islands).

[14]See http://sasi.group.shef.ac.uk/worldmapper/posters/worldmapper_map295_ver5.pdf Map of CO_2 emissions (Source: SASI Group, University of Sheffield; Mark Newman, University of Michigan, 2006 (updated in 2008), www.worldmapper.org. Source of data: Gregg Marland, Tom Boden and Bob Andres, Oak Ridge National Laboratory).

cities is widely debated. Many authors assert cities are the main contributors to greenhouse gases, and even more so as regards high-income cities (Satterthwaite 2008, Romero-Lankao 2007; Dodman 2009). Nevertheless, in the determination of the spatial differences in emissions, identification of area of attribution is fundamental, and this often does not correspond to administrative (municipalities, metropolitan areas, areas of continued construction) or functional boundaries (city-region) (Davoudi et al. 2009). The main sources of greenhouse gas (energy production stations, garbage dumps, big industry and large transportation hubs) are often outside urban limits, and even if they are not directly attributed to the city, they are probably linked to systems of urban consumption. For this reason, some authors (Satterthwaite 2008) highlight that emissions of anthropic origin are linked to the consumption models of medium- to high-income groups, and it is therefore necessary to address these in order to implement mitigation strategies.

4.2.2.2 The City as a Victim of Change

The other side of the changes that occur in urban environments is constituted by the impacts that they have on health, biodiversity and people's quality of life. The 2007 IPCC report reflects the growing risk of extreme climate events (storms, flooding, drought) and environmental stress due to changes in rainfall patterns, temperature and air humidity (see Appendix V). Like the causes of changes, the intensity of their impacts on health, biodiversity and people's quality of life will be different depending on the geographical, environmental, economic and social conditions that influence vulnerability in a given context. Low-income social groups, who have limited resources at their disposal, have more difficulty insuring their assets or guaranteeing access to water, electricity, sanitary services, sewers and other basic services (for example in terms of moving to safer areas) (Davoudi 2009; Satterthwaite 2007). The range of environmental, economic and social consequences that climate change may produce in cities, and in particular in low- to medium-income countries, will depend largely on the manner in which they are organized and managed, from their fiscal balance, to the quality of their infrastructure, to their method of managing resources (Satterthwaite 2007). UN-Habitat has asserted that the urban dimension of climate change still has a limited presence in the international debate, and that since the impacts of climate change compromise countries' efforts to reach sustainable development objectives, adaptation is necessary. This is fundamental, especially because the climate is already changing and the majority of countries do not have the adaptive capacity to respond to or address the impacts of climate change on cities, settlements and livelihoods (UN-Habitat 2008).

Attention to adaptation measures for climate change effects in the cities of low- and medium-income countries is necessitated, in addition to the above-mentioned vulnerability factors, by the fact that they represent nearly three quarters of world population, and will be the site of most of the population growth that will occur over the next few years, casting them in a determinative role as regards the production of greenhouse gasses. Their population, which grows at an elevated rate, is highly exposed to risks of storms, flooding, and other extreme events and environmental stresses linked to climate change (UN-Habitat 2008). Environmental changes can exacerbate already critical hygiene situations, promoting the spread of diseases such as malaria (Wilbanks et al. 2001; Parry et al. 2007). The rapid informal expansion of cities without health care infrastructure and in areas of environmental risk (e.g. hydro-geological risk) could combine with environmental transformation, intensifying the impacts of climate change, and forcing local and national governments to dedicate extensive resources to the reduction of urban vulnerability. For these reasons, even in the sphere of urban studies (Bulkeley 2006; Byrne/Jinjun 2009; Gleeson 2008; Newman et al. 2009; Smith et al. 2010; Wilson/Piper 2010 cited in Matthews 2011) it is strongly maintained that there is an urgent need to adapt urban systems to climate change, and therefore to develop specific studies and policies for the adaptation of human settlements to future climate changes in order to minimize the risks they constitute in terms of people's well-being (Gleeson 2008).

Some development studies have raised important questions on the adaptive capacity of populations and communities, highlighting the relationship between the lowest levels of adaptive capacity and poverty (Dow et al. 2006), and therefore how the question of social justice is implicit in adaptive capacity. In general, medium- to low-income countries have limited adaptive capacity because their financial resources are limited and their institutions lack the capacity to mobilize them. The justice issue is amplified by the fact that the impacts of climate change are probably linked to processes of industrialization in developed countries, and thus to the emissions associated with decades of economic growth in wealthy countries, while the communities most impacted by climate change are located primarily in the poorest areas that have not benefitted from that growth.

As with other complex and transversal issues that have broad social and economic implications, government intervention is considered most effective if implemented through the integration of adaptation to climate change within policies for spatial planning and for development more generally (Klein 2003). This integration should not be understood as the introduction of specific adaptation measures into the planning and implementation of territorial development strategies, but rather as mainstreaming, i.e. the consideration of adaptation to climate change in the decision-making and planning processes. The advantages of mainstreaming include guaranteeing long-term sustainability of investments and reduction of sensitivity to environmental change (Hug et al. 2003; Agrawala 2005; Klein et al. 2005; Eriksen et al. 2007). Mainstreaming is therefore viewed as a strategy for achieving more effective and efficient use of financial and human resources if compared to the definition, implementation and management of climate policies in a manner that is separate and independent from the development initiatives already underway.

Despite the emphasis placed on adaptation following the Bali conference, there is still an imbalance between projects and programmes on mitigation strategies and those for adaptation. In fact, the majority of financing is directed to mitigation in both high-income countries and in LDCs. Moreover, most projects and programmes for cities are focused on reducing their contribution to climate change, while those on adaptation are found predominantly in rural and agricultural areas, particularly in the poorer countries, and concentrate on livelihoods and exploitation of natural resources that are sensitive to climate change. Finally, the existence of different financing streams for mitigation and adaptation contribute to keeping the two strategies separate and limit the development of projects that involve a synergic interaction between the two.

4.2.3 *Approaches and Strategies for Adaptation in Urban Areas: Oxymorons and Opportunities*

Notwithstanding uncertainties regarding the causes and impacts of climate change, there is a shared recognition of the implications that spatial configuration of the city and the way land is used and urbanized have in terms of reducing the emissions causing change and adapting to their impacts (Davoudi et al. 2009). The form of settlements, their impact on natural resources and their emissions level are all influenced by a variety of complex factors, including the constructive technology at their disposition, the land and housing market, investment strategies of private and public institutions, public policy (regarding, for example, planning, housing development, transportation, environment and the fiscal system), the institutional tradition, social and cultural norms, individual lifestyles and social behaviours (Davoudi et al. 2009).

On the other hand, the objectives and the contents of spatial planning are subject to a variety of interpretations. In this study, and pursuant to the assertions of Davoudi et al. (2009: 1), spatial planning is understood as a "critical thinking about space and place as the basis for action or intervention (RTPI 2003: 2)", which extends beyond the normative framework for land use, and includes institutional and social resources through which the framework is applied, challenged and transformed. In this sense, spatial planning is conceived as a process for resolving the problems of a specific place, and its objective is sustainable development (Davoudi et al. 2009: 13–15).

4.2.3.1 Spatial Planning as an Agent of Adaptation

It has been asserted that cities play a key role in the effectiveness of adaptation and mitigation policies. Many initiatives have been developed in recent years in order to reduce the contribution of cities to climate change and to mitigate the impacts of the transformations under way (e.g. the C40 Climate Leadership Group at the international level and the Majors Alliance for Climate Change Protection in the U.S.). Nevertheless, urban areas have historically initiated activities to reduce their sensitivity to climate changes and variations, for example by adapting the buildings in which people live or work, controlling the flows of canals and rivers, or modifying the configuration of their coastline. Recent climate changes, however, pose challenges that seriously test the capacities that have been used so far (Romero-Lankao 2008).

Two main mechanisms are used to address climate risks: risk management and adaptation strategies.

The former has a long history in the political and scholarly sphere, in which consolidated models for understanding and managing modalities of response to environmental disasters have been developed. This approach provides for the development of a series of risk management actions throughout the entire cycle of the disaster: prevention actions (also know as mitigation actions) and preparation, emergency response actions, and reconstruction and recovery.

Adaptation has a shorter history, which is mostly linked to research and action on climate change. In the urban sphere, it is essentially defined as the construction of the city's adaptive capacity (or resilience), understood as 'the potential' of a city, its population and its political decision-makers to modify the physical-functional urban balance and/or people's behaviour in order to better address existing and predicted climatic stimuli (Romero-Lankao 2008: 64). Which factors determine adaptive capacity and how it is constructed are still topics of open debate, which is focused primarily on its relationship with other approaches to vulnerability and resilience (see Chap. 5) that orients intervention towards increasing the resilience of the urban system and/or reducing the vulnerability of infrastructure and people.

The processes of urban development and spatial planning may improve or limit the adaptive capacity of a city's inhabitants, especially low-income groups, thus determining the potential for effective adaptation (Romero-Lankao 2008).

Adaptation is heavily determined by the availability of resources, such as knowledge, infrastructure, the quality of institutions, governance systems, and technological and financial resources, all of which create opportunities to initiate adaptation options in various sectors (Romero-Lankao 2008; see Appendix VI).

Institutions play a key role in improving adaptation in urban areas, even though collective (e.g. community-based adaptation) and individual capacity also have a fundamental role in addressing environmental change, especially in informal settlements (Stephens et al. 1996; Action Aid International 2006; Dodman et al. 2010, cited in UN-Habitat 2011; Satterthwaite et al. 2007).

The role of local and national institutions is crucial for two reasons: first, the control exercised by systems of land use management are at the heart of vulnerability since populations located in risk areas often depend on them; and second, in terms of urban development, it is the prerogative of institutions to ensure that inhabitants have access to infrastructure and services. Moreover, it is increasingly recognized that to effectively address climate change an integrated action at multiple levels of government and within the spheres of politics, economics and society is require (Albert/Kern 2008).

For institutions to include considerations on climate change in urban planning means guaranteeing that construction of infrastructure be cognizant of climate risks, guaranteeing broader and more appropriate access to information on climate change and on local impacts, and agreeing to support strategies and programmes to avoid or prepare for environmental disasters. Adapting the infrastructure that provides social, environmental and economic[15] services to climate risks requires a rethinking of priorities and therefore of investments. Insurance is one support strategy that is increasingly used for adaptation, especially in high-income countries but also in those that are medium- and low-income. This strategy seeks to reduce losses caused by climate events (IPCC 2001, 2007) through new insurance mechanisms that are defined in such a way that better distributes losses, for example by expanding insurance coverage on property, and can also act as an incentive to reduce risk through adequate construction regulation and flood prevention plans.

The development of new infrastructure and the adoption of new finance and insurance mechanisms are certainly two types of measures that are prioritized in the majority of strategies currently being implemented, and in the ongoing debate on adaptation. The type of infrastructure to use, however, is a topic of frequent debate, and some authors who are critical of the dominant 'securitization' of the city approaches (Marvin/Hodson 2009) assert that they are too oriented by western development models and lifestyles, and necessitate a technology transfer and solutions that are independent of the context under consideration, rendering them risky and unsustainable.

4.2.3.2 Urban Ecological Security: A Dominant Approach?

In practice, dominant approaches seem to be fundamentally traditional (Simon 2003) in both spatial planning and the choice of adaptation options, giving limited consideration to more recent post-structuralist approaches that are centred on people, pluralistic, participatory and oriented to sustainable economic, social and environmental development (Simon 1999, 2002, 2003; Adams 2000; Long 2001; Nederveen Pieterse 2001; Reed 2002, cited in Simon 2003). This constitutes, for some, one of the reasons for the lack of improvement in the quality of life and empowerment of many people in African cities, causing the failure of regional and local initiatives that simulate development, while in reality serving the economic interests of a national and regional elite (Simon 2003).

On the other hand, new approaches are making progress through policy mobility (Cochrane/Ward 2012) and by linking closely with political and global environmental issues. The environmental conditions (climate change and limited resources) and policies (the spectre of terrorism since 9/11/2001) of recent years have in fact led to a growing interest in human and ecological 'security'.

The term 'ecological security' has traditionally been used in relation to attempts to safeguard the flows of ecological resources, infrastructure and services at the

[15]Such services include flood control, water provisioning, drainage, management of solid, liquid and hazardous waste, energy, transportation, other urban development work, residential areas, commercial and industrial activities, and recreational areas (Kirshern et al. 2007, cited in Romero-Lankao 2008).

national level. Recently, however, it is increasingly associated with concerns related to Urban Environmental Security (UES) through forms of *Secure Urbanism and Resilient Infrastructure* (SURI). Such approaches are configured through strategies that aim to defend the city and its infrastructure in order to guarantee ecological and material reproduction. The main problems that UES seeks to address include the limited nature of resources, and climate change (Hodson/Marvin 2009). 'Strategic protection' against the impacts of climate change (specific long-term strategies), 'autarchy' or self-sufficiency in the provisioning of resources, goods and services, the construction of urban 'agglomerations' with efficient and closed systems for energy and transportation, are the three principles upon which UES is based (Hodson/Marvin 2009).

The approaches developed along these lines propose the 'global city' as a model for the development of policies, security, infrastructure and governance that integrate ecological and economic aspects while supporting a relationship between man and nature that is oriented to economic growth (from a competitive to an eco-competitive city, from sustainable cities to Urban Ecological Security, from infrastructural vulnerability to strategic resilience) (Hodson/Marvin 2009). These kinds of approaches, applied indiscriminately, raise several issues that question the legitimacy and effectiveness of their replication in contexts such as the sub-Saharan African city. Processes of spatial valorization (e.g. privatization or the development of infrastructure and new technology) linked to such approaches often represent a new form of what Harvey (2001) defines as a 'spatial fix' (an instrument for remedying the over-accumulation crisis through investment in new sites of value). Three regulatory theories exist which are again based on western urban experiences, whose shortcoming is that they ignore some of the main methods of the production of space (Lefebvre 1991) in other urban and metropolitan contexts, such as sub-Saharan Africa. The extent to which this new 'logic' is at the basis of new strategies of economic accumulation or more 'progressive' policies has been the question that Marvin and Hodson have sought to answer in their analysis of global level strategies.

Independently of the basically explicit objective of these strategies, it is certainly important to understand what their impact is on the capacity of people and institutions to face the environmental changes underway. Do they still represent a valid instrument to help people and institutions improve their adaptive capacity, anywhere in the world?

Moreover, these new approaches lead one to pose questions regarding several key aspects: firstly, whether they signify the passage from competitive relationships to eco-competitive relationships between cities as regards governance and urban economy; secondly, whether urban 'securitization' and the resilience of infrastructure reflect a centring of the conventional debate on sustainable cities around 'urban environmental security'; and finally, whether the approaches adopted (SURI) signal a shift towards resilience as a strategy for addressing infrastructural vulnerability.

It is therefore necessary to critically evaluate the implications of this new logic, which according to Marvin/Hodson (2009) can be summarized into three groups.

First, we are witnessing a growing 'metropolitanization' of resource security and the responses to climate change, which leads to strategic relocation and 'selective' global urbanization with respect to ecological resources. World cities use their capacity, resources and networks to overcome the potential limits of climate change and resources in order to guarantee future economic and territorial growth. Inquiry into what this means for all the by-passed localities, new peripheries, and 'ordinary cities' in the North and South is essential. The implication is that cities simply act or 'improvise' with their limited resources and capacity, and construct ecologically secure spaces. On the other hand, 'ordinary cities' (see Chap. 2) and the cities of the South are viewed as potential new markets that 'consume' eco-city type architectural and engineering solutions produced in exemplary western cities.

Second, longer timelines can be blocked into socio-technological trajectories and lead to the formation of social, economic and spatial plans in which priorities are principally defined in terms of protecting the interests of western cities. The risk is that only certain technological knowledge will be prioritized, and that the spectrum of what is considered relevant knowledge and experience is too limited. More generally, this concern is related to the limitation (if not the outright exclusion) of groups representing diverse social interests and involved in the construction of new change horizons, and above all the indiscriminate replication of new eco-models in different national and international contexts (Marvin/Hodson 2009).

Coalitions and 'trans-urban networks' of city government, and social and environmental groups are emerging to produce 'new solutions'. The political elite of large cities, society and environmental groups are positioning themselves as actors and places to confront the 'threat' of limited resources and climate change. There seem to be coalitions of social interest groups, apparently very exclusive, which are involved in the development of an agenda that claims to speak on behalf of everyone, as was also reflected by the conflicts that arose during the recent Conferences of the Parties on climate change. One wonders what the benefits are of urban development strategies that are decided at the global level, and who, where and what they may impact.

The risk is that material and economic interests will prevail, causing the development of new coalitions of cities, governments, enterprises and environmental associations in order to develop actions under the name of Urban Environmental Security, actions that may turn out to be designed to protect the economic interests of private parties and the few.

In order to avoid these risks, it is essential that new urban forms and styles be tested and launched in order to transcend the conventional notions of infrastructural and environmental limits, in contrast with the construction of systems of neutral resources or of more autonomous cities that redesign and substitute interrelations and resource dependence. The path indicated by these approaches runs the risks of excluding alternatives, such as those that can be created in the peri-urban context of sub-Saharan cities, and of amplifying political tensions, conflicts and social resistance to such approaches if they are imposed. There are alternative approaches to this dominant logic, transition towns for example, but they have limited visibility and are relatively unexplored. There should be a greater knowledge of the specific context at the centre of interpretive approaches, led by an interpretive approach to the African city that is not focused only on the gaps and the absence of western 'urban' elements.

Analysis of the dominant approaches under consideration reveals that they push for increased economic competition between the neoliberal cities of contemporary global capitalism, which invests resources in the 'race' to secure cities and strengthen economic growth. In cases where the main strategic resources are vulnerable, and those same resources (at the basis of economic competition between cities) also support the material, social and ecological reproduction of a city, conflicts can arise between reproduction of the city and economic competition, which is based on divergent interpretations of the notions of 'ecology' and 'security'.

4.3 Planning Practices in Dar es Salaam

Allen Armstrong, in his analysis of colonial and neo-colonial urban planning in Dar es Salaam, maintains that the growth of a city "occurs parallel to, and been strongly influenced by, the wider development of urban planning as a modern practising discipline" and that "despite the continuous transformation or even revolution in local conditions and wider national policy to which each plan had to address itself, nevertheless, the strongest impact on each plan has been Western planning values and concepts in general and, the planning fashion prevailing at the time of each plan's preparation in particular" (Armstrong 1986: 43–44).

The first general schema for city planning dates back to the period of German colonialism (1891–1916), when Dar es Salaam became the capital of German East Africa. At that time, what was, and still is, the commercial and economic centre of the city to the north of the natural port was designed. The approach adopted was that of urban design, which would continue to inform the planning of Dar until the Second World War, providing considerable continuity between German domination and the subsequent British colonization (1918–1961). A new approach, in line with the master planning tradition, was introduced with the second schema for city planning, developed after the Second World War.

Both plans, notwithstanding the revisions made in 1968 and 1979 during the post-independence period, proved to be a failure. The colonial schema is paradoxically reproduced in the new versions developed by the Government of President Nyerere: the zoning and development provisions for the city centre remained unchanged, the elimination of informal settlements through the transfer of inhabitants to new legally registered lots, the control of the population's movements and the densification of the city all remained. Nearly all the provisions made during those years were distant from the real development of the city, and were never implemented (Halla 2007).

Friedman (2005) attributes the ineffectiveness of the first urban plans prepared by the African government after independence to a combination of factors: the inadequacy of the approaches and instruments used, the incapacity to address urban dynamics (high population growth rates and reduction of financial resources), and blindness to the structural weakness of the public administration in these respects. This is fairly normal if one considers that these plans were often financed by international donors, and their development, entrusted to western consultants, occurred elsewhere and according to a logic that was completely foreign to the context to which they were applied.

With the economic reform of 1985 imposed by the Structural Adjustment Programmes of the World Bank, urban governance opened up to public-private partnership and planning came to be informed by "coordinated and decentralized" objectives, or "urban management" (Halla 2007: 134). The programmes initiated in the 1990s by the United Nations introduced the Environmental Planning Management strategy in order to facilitate the emergence of a local planning culture, which would be defined according to priorities (natural resources as a fundamental element for the livelihoods of the majority of the urban population), the uniqueness of the settlement system (consisting mostly of vast peri-urban areas with hybrid rural–urban characteristics and different from both urban and rural areas) and by existing social and institutional capital (a complex product of the interaction between tradition and modernity). At the centre of this new culture, the idea of action planning was recognized (Friedman 2005) as a planning modality that was heavily oriented to comprehensive intervention, which included all the strategic themes of urban development and sustainability, while also addressing the social, economic and environmental dimensions of development.

4.3.1 Master Plans in the Dar es Salaam Region

4.3.1.1 The 1949 Master Plan

Defined in the optimistic context of colonial development and according to the welfare scheme adopted by the British in the decade 1947–1957, the 1949 Master Plan delineated a rational model of growth for the city viewed as the major administrative and commercial centre of Tanganyika.

The influence of western planning approaches is evident in both the application of the Garden City principles and in the choice to concentrate on a series of problems at the heart of British planners' concerns at the time: health and aesthetics, as well as the provision of open spaces and the containment of urban sprawl. These principles and priorities combined with the desire to maintain strict racial segregation in the spatial distribution of the population. To that end, the plan introduced a series of imbalances in the distribution of services between various areas, favouring the small elite comprised of the colonial community employed in administration and business (Armstrong 1986).

Residential areas were subdivided into zones with services and low density for the Europeans, zones with medium density for the Asians, and zones with higher density level for the Africans. The plan paid disproportionate attention to the planning of the low-density neighbourhoods for Europeans (between 2.5 and 6.1 % of the total inhabitants). The proposals for the areas in which the Africans would reside were limited to reducing density, assuring that the construction of houses respected the road network, and providing a common water source. The concession of lots for agricultural activities was also planned, and native forms of settlement, including the *shamba* (cultivated field) could be developed, similar to the gardens cultivated by industrial workers in England and other European countries in the 1920s and 1930s. The plan recognized the settlement of many Africans in peri-urban areas, organized according to traditional modalities to which elevated construction standards could not be applied (Schmetzer 1982, cited in Šliužas 2004: 85).

The plan also provided for zones intended for industrial use, implicitly favouring the modern capitalist sector that required a wide array of services and easy access, and discriminating against the intense artisanal work carried out at home, according to local tradition, which was practiced by the majority of Africans who sought to improve their condition through a mix of activities and land uses.

The green areas and bodies of water that characterized the natural environment of Dar es Salaam were seen in the plan as elements to conserve in order to guarantee good sanitary and environmental conditions; where these natural structures were absent or altered, they were created artificially in order to allow for good ventilation of the city (breeze lines, ventilation funnels) through continuous corridors of open space that ran from the coast to the inland built-up areas along the main lines of wind flow.

Although attention was paid to natural environment, the environmental management practices already adopted by Africans in urban and peri-urban areas, where not considered as an opportunity to protect health and preserve environment and resources. Various urban problems were addressed in purely physical terms to consolidate the pre-existing urban structure and applying western urban models. Overall, the plan therefore had a limited impact on the transformation of the city, which continued, particularly outside the planned schema.

4.3.1.2 The 1968 Master Plan

In the mid-1960s, recent independence and the role of Dar es Salaam as the capital of the new state determined profound changes in the conditions of the city. The growth of the population in those years exceeded forecasts by 35 %, and 70 % of the population lived in informally occupied areas. The new Master Plan was heavily influenced by the new political orientations of the government, ratified by the Arusha Declaration, which marked the birth of Tanzania as a socialist state and nationalization of land, industry and infrastructure.

The 1968 Plan, returning to the notion of urban systems, sought to develop future models of the city on the basis of a rational system integrated with land use and transportation. The city was conceived as a combination of base cells with a 'nested' hierarchy within which sub-urban residential expansion was organized, with satellite 'cities' that were comprised of 4–5 residential communities or neighbourhood units.[16]

Unlike the previous plan, the 1968 version made an important effort to incorporate the dominant political and social preferences of Tanzanian society and tried to break down the barriers of racial discrimination by favouring a variety of lots and building typologies, reducing settlement density in overcrowded areas (e.g. Kariakoo), mainly in zones where Africans resided, and introduced new low-density zones (e.g. Oyster Bay). The plan also paid attention to the question of implementation, including specific practical measures, an investment programme for public works, and plan monitoring, which was necessary during the first five years in order to translate the plan into a reality. Awareness of the importance of instituting a permanent planning and monitoring unit lead to the creation of a section for the Master Plan in the administrative offices of the city.

Despite these innovative contents, there was a substantial continuity with several elements of the previous plan still persisted (Armstrong 1986; Šliužas 2004).

One such element was non-functional zoning, which was at the basis of spatial population segregation processes, and which remained as an unchallenged characteristic. The plan allowed for industrial settlement in appropriate areas that were close to the residences of workers in order to reduce their commute, but in general the principle of segregation was reproduced in the distinction between four different classes of industrial zones and the introduction of sub-classifications for areas with other functions.

The plan is criticized, mainly because it addressed the issue of informal settlements, declaring them illegal (squats) and in an authoritative way, paying little attention to the needs of informal dwellers. Those settlements were considered as a problem for the development of the city, and were therefore to be prevented or eliminated, reflecting a hostile and intransigent attitude similar to that which had been adopted by many urban administrations of the time. The 1968 plan provided for the elimination of all squats by 1990 through an integrated combination of measures that included the removal of existing settlements, the development of new residential neighbourhoods, increases to staff in planning offices and zero compensation for the costs of resettlement or the inconveniences caused by removal activities. This approach came up against higher demographic growth than had been forecast, which together with continued migration to the urban area quickly led to the expansion of informal areas in which the majority of city residents lived, rendering the coercive measures that had been proposed both administratively and politically impracticable. Moreover, the proposed interventions were over-ambitious in terms of the actual availability of funds and administrative capacity (Armstrong 1986).

---

[16]A modified concept with respect to the interpretation of the 1948 plan defined larger communities of about 40,000 inhabitants, which contained a secondary school, a large central market and small-scale industrial activities.

4.3.1.3 The 1979 Master Plan

The 1979 Master Plan (see Fig. 4.1) was developed during a critical period characterized by the proliferation of informal settlements, unemployment and deterioration of environmental and health conditions. Furthermore, Dar es Salaam was involved in a series of political transformations, including: relocation of the capital to Dodoma and administrative decentralization; a radical revision of housing policies that accepted the presence of informal unplanned settlements; a shift in the *National Housing Corporation's* intervention priorities from elimination of slums towards the construction of dwellings and the provision of basic services; and finally, the reorganization of local government with the transfer of local urban responsibilities from *City Council* to new regional authorities.

The new plan seemed to be more adapted to the territory and a pragmatic and flexible document. The old 'rationality had given way to the politics of group conflict, technology to discussion, long-term planning to short-term management, promising immediate solutions to pressing problems' (de Tiran cited in Armstrong 1986: 58).

Six alternative schemas were delineated during the preparation of the plan, and chosen only after consultation with external consultants from the *Ministry of Land and the City Urban Planning Committee*. On the basis of flexible provisions for

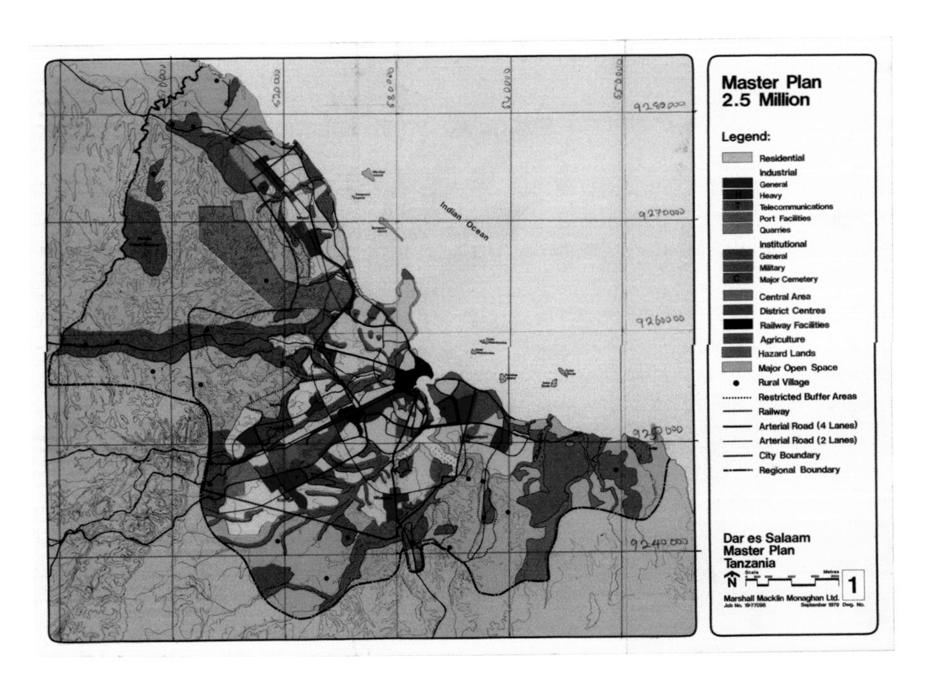

Fig. 4.1 Dar es Salaam Master Plan 1979. *Source* (Marshall Macklin Monaghan 1979)

population growth, three development phases of the city were identified and linked to three difference population scenarios (1.2, 1.5, or 2.4 million people). The plan was a document that would serve as a dynamic guide to monitoring the growth of the city, to adapt to different and changing circumstances, to address conditions of uncertainty with an increasing number of variable and an expanding temporal horizon (Armstrong 1968: 58), without references to stereotypical concepts such as satellite cities, green belts and planning units.

Moreover, the objectives of elimination and demolition of slums in previous plans were substituted with those of conservation and gradual improvement, also because informal settlements housed the majority of the population, which now had rights recognized by the government. Although new development was to occur in surveyed and registered plots, the self-construction of dwellings with local materials was subsidized. This favoured the no longer illegal development of areas with mixed rural–urban characters, and the self-constructed dwelling was often associated with cultivated areas for the production of food and/or income.

In addition, interventions were planned for monitoring without interfering in the development of new informal settlements. In fact, future unplanned development was admitted in the concept of 'residential buffer zones', which represented a new and interesting concept for the inclusion of further dwellings in areas where there was strong pressure from the population.[17] These were "areas adjacent to short and mid-term development areas but not on land designated for future development, but where no facilities will be provided and where the installation of utilities will become solely the owner's responsibility, making them, effectively, officially recognised future squatter areas and institutionalising the dualist concept of two classes of urban resident" (Armstrong 1986: 61).

Another innovative element of the plan was the introduction of participation, through the involvement of politicians, administrators and representatives of the most important national institutions in the formulation of proposals and initial orientations.

Notwithstanding the introduction of these innovative elements, the urban structure proposed was substantially similar to the previous plan, planning continued to focus predominantly on the physical aspects of the city, the movements of people in the territory and the reduction of density by preventing immigration. In addition, despite the efforts made to incorporate flexibility and pragmatism, the 1979 Plan was hampered by limited resources, reduced capacity for implementation and the pressures of demographic growth.

[17]This favoured, however, the production of informal areas with critical living and environmental conditions, which led for example to the obstruction of the valleys of certain waterways, changing the natural drainage conditions and contributing to flooding and the creation of health risks.

4.3.1.4 The New 2012–2032 Master Plan

The new plan for Dar es Salaam (still unapproved as of April 2015) has been presented in several events and political directions have emerged from the dialogue with local and national institutions (City Union of Dar es Salaam, Officer of the MLHHSD). In February of 2011 in the offices of the *Ministry of Land, Housing and Human Settlements Development* (MLHHD) in Dar es Salaam, an agreement was reached with an International Consortium of the Permanent Secretary of the MLHHSD, which agreed to consult on the preparation of the New Master Plan for Dar es Salaam City 2012–2032 (Dodi 2011).

That agreement is the final result of a competitive tender organized in 2009 by the Ministry with the support of the World Bank and other donor agencies. The consortium consists of the combination of international and local organizations.[18] The project will move forward with the collaboration of a local technical team composed of personnel from the Ministry, from the Dar es Salaam City Council, and from the three municipalities (Ilala, Kinondoni and Temeke).

Within the Master Plan, ambitious objectives have been established, including the provision of a plan for the future urban development of Dar es Salaam, not only as a city leader in Tanzania, but also as an important metropolis in all of Eastern Africa. At the same time, the plan should do address several critical questions that concern the city (for example, traffic and transport, infrastructure, informal settlements, etc.) (Dodi 2011).

The Plan aims to create an 'inclusive' habitat and promote a balanced spatial growth and a functional organization, by developing a network of new centralities, in both the existing formal and informal city and in the future city, and a new structure of public spaces. The Master Plan has developed also several proposals (protect natural areas, develop a green infrastructure strategy, limit coastal development. Designation of marine or coastal protection areas) for the improvement of urban and environmental quality, paying attention also to climate change and the adaptation practice (Fontanari forthcoming).

The plan will need to offer opportunities to build the capacity of a team of local experts that will facilitate the effective implementation of the plan, and will be an instrument for good future urban management of Dar es Salaam.

The representatives of the International Consortium have worked to provide a critical assessment of the history of the Dar es Salaam Master Plans (Table 4.1), including the 1979 plan, which basically was never implemented, and the draft of the Strategic Plan in 2005, in light of the new *Urban Planning Act* (2007) which

[18]The consortium is comprised of four organizations: the Italian Dodi Moss LLC, Buro Happold Ltd out of London, as well as Q Consult Ltd. and Afri-Arch Associates, both of which specialize in local practices. These companies represent a team of international experts that have worked in similar roles throughout the world, from Africa to Europe, America, and Asia.

Table 4.1 Summary framework of the Master Plans of Dar es Salaam

Year of publication	Title and content	Consultants	Funding	Some of the major planning concepts
1949	A Plan for Dar es Salaam (158 pp + 12 pp of appendices)	Alexander Gibbs and Partners, London (1947–1949)	Britain	Zoning of functions
				Zoning of residential areas according to density and race
				Neighbourhood Unit 'Breeze lane', open spaces provision
				Non-geometric street layouts
				Density and building standards
1968	National Capital Master Plan: Dar es Salaam Main Plan Report (157 pp + 7 pp of technical appendices)	Project Planning Associates Canada Ltd, Toronto (Jan. 1967–Feb. 1968)	Canada	Plan 2000 (long range concept)
				Systems Approach
				Ecosystem of growth/hierarchical modular urban structure including neighbourhood units, satellite sub-cities and city-region planning
				Green belt, parkways, landscape corridors, open space provision, sector strategies.
	Capital Work Programme (57 pp)			Five-Year Capital Works programme
1979	Dar es Salaam Master Plan: Summary Main Report (104 pp)	Marshall. Macklin, Monaghan Sweden Ltd., Toronto (Nov. 1977–April 1979)	Sweden	Flexibility—population attained rather than target years
				Hierarchical urban structure based on planning module
				Sub-classification of residential areas/recognition of squatter areas
				Participation of implementing agencies
	Five-Year Development Programme (60 pp)			Detailed implementation programme including 47 priority projects
	Technical Supplements			

(continued)

Table 4.1 (continued)

Year of publication	Title and content	Consultants	Funding	Some of the major planning concepts
2020–2030 (not yet published)	New Master Plan for Dar es Salaam City 2012–2032 (Dodi 2011)	International Consortium: Dodi Moss LLC (Milan), Burro Happed Ltd. (London), Q Consult Ltd., and Afri-Arch Associates (Dar es Salaam)	Italy, Great Britain and Tanzania	Strategic objectives: Make Dar es Salaam a city leader in Tanzania and a central metropolis for all of Eastern Africa
				Address the main critical issues, including traffic and transport, infrastructure and informal settlements

Source Adapted from Armstrong (1986: 49)

delineated the guidelines for the principles and practices of planning for the new century, and have already outlined the technical approach and the methodology for the development of the new Master Plan (Dodi 2011).[19]

4.3.2 *Action Planning and Participatory Planning*

At the beginning of the 1980s, flexible planning seemed to have been replaced by a new approach oriented to action planning (Honeybone 1978 cited in Armstrong 1986) in the short term, and focused on management of the crises underway, which afforded only minimal consideration to the directions and prescriptions in the existing Master Plan. Nevertheless, the conditions in Dar es Salaam continued to worsen during the 1990s, after years of economic difficulty and inadequate management by institutions, the planning reform in Tanzania began. For the city of Dar, this meant the beginning of the development of the fourth Master Plan, for which the *Ministry of Land, Housing and Urban Development* (MLHUD) requested financial and technical support from international donors. The failures of previous

[19]The following phases are planned: (1) Background analysis; (2) Analysis of the context (Dar es Salaam in the local, national, regional and international context); (3) limits and opportunities; (4) Planning philosophy; (5) The strategic planning process; (6) Strategy/approach to sustainability; (7) Working in partnership with stakeholders; (8) practicable financial projects and programs—public-private partnership; and (9) Work plan: capacity building for good implementation of the Master Plan. The communication process is understood as a work in progress, open and directed at suggesting and receiving contributions in order to define "the vision and objectives of Dar es Salaam as a new African metropolis" (Dodi 2011).

planning experiences led to the definition of a new method of planning and urban management, which was concretely applied through the *Sustainable Dar es Salaam Project* (SDP) developed with the support of the *Sustainable Cities Programme*[20] (SCP). The SDP generated a *Strategic Urban Development Planning Framework* (SUDPF), completed in 1999, which constitutes a general framework for the management of the physical development of the city.

The idea of city planning in parts and the development of prescriptions and projects for individual neighbourhoods was abandoned in the SDP, which opted for the formulation of a series of guidelines for the entire city intended to orient the upgrading of existing and future urban expansion. Through consultation with different stakeholders, priority areas of intervention were identified and thematic working groups were defined that would concentrate on the development of actions and exemplary projects with respect to several key questions for the planning process: support for city expansion, upgrading of settlements that lack services, management of solid refuse, management of surface water and sewage, management of traffic and atmospheric pollution control, management of open spaces, risk areas and urban agriculture, management of informal commerce, management of urban renewal and management of construction materials and coastal resources (SDP 1999).

Three types of strategic orientation inform the SDP: spatial, managerial and financial. From the spatial perspective, the decision was made to identify the ecologically sensitive areas and those potentially suitable for urban development on the basis of three criteria: environmental risks, the cost of developing infrastructure and economic efficiency of investments. In terms of management, the SDP sought to develop forms of public–private partnership and to directly involve settlement communities. Finally, as regards finance, the objective pursued was that of facilitating the generation of resources in order to render investments sustainable.

Through the *Environmental Planning and Management* (EPM) process, the SUDPF became a new instrument to capitalize on local resources and developing a series of upgrading programmes for the existing city that provided for the direct participation of inhabitants in the development of infrastructure.[21]

The entire national reform of the urban management and planning system, whose first step was the Human Settlement Development Policy of 2000, is positioned along the same lines. This policy document established that, rather than demolishing unplanned settlements local governments would promote their upgrading

[20]The Sustainable Cities Programme is a UN-Habitat/UNEP program initiated in the early 1990s to support the planning process in cities and to help them achieve more sustainable growth in terms of environment and development. It is based on a broad process of participatory decision-making in the urban sphere, and promotes the sustainability of cities through a strategy of Environmental Planning Management (EPM).

[21]Community Infrastructure Programme (CIP) and Community Infrastructure Upgrading Programme (CIUP).

through participation. This strategy is at the basis of the 2010 *Citywide Action Plan* (CAP), whose task was to further the upgrading strategy for informal settlements lacking infrastructure and services, launched in 2007, through the distribution of regularly registered plots, the provision of basic services and improvement of settlements.[22] The objective of the CAP is to upgrade 50 % of the unplanned settlements lacking infrastructure and services by 2020, and to prevent the formation of new ones through action on three principle components: land, basic services and settlements. With respect to the first component, the plan seeks to guarantee access to planned areas through the 'regularization' of unplanned areas, to increase revenues deriving from taxes on land, to increase the number of new planned lots that are economically accessible and to control densification of residential areas (UN-Habitat 2010). One characteristic of the Citywide Action Plan is that it coordinates with all the plans and projects underway in the development of the city, particularly with the New Master Plan. The most important instruments are listed below, and grouped according to sphere of intervention.

(a) Upgrading of informal settlements and regularization of land access:

- Community Infrastructure Upgrading Programme (CIUP)
- Formalization of unplanned areas through Residential Licenses
- 20,000 Plots Project
- Provision of planned land to low-income households through a community-based approach in Chamazi
- UN-Habitat-supported Tanzania Financial Services for Underserved Settlements (TAFSUS)

(b) Upgrading of central areas and development of new urban centres:

- Kurasini Redevelopment Project
- Satellite Cities Project
- Kigamboni New City Project

(c) Expansion of infrastructure for mobility:

- Traffic decongestion through Dar Rapid Transport Project (DART)
- Expansion of Julius Nyerere International Airport
- Dar es Salaam road expansion programme

[22]Another goal of the plan was to contribute to achieving the development objectives contained in a series of national policy documents (such as the National Growth and Poverty Reduction Strategy—MKUKUTA, the National Vision 2025, the National Housing Policy and the National Human Settlements Development Policy) and international policy documents (the Habitat Agenda and the Millennium Declaration, particularly as regards MDG 7, Target 7°: water, sanitation and slum upgrading).

(d) Expansion of hydro and sewer networks:

- Dar es Salaam Water Supply and Sanitation Project (DWSSP);

(e) New industrial settlements:

- Kigamboni oil refinery project.

Although the 1979 Plan initiated the process of questioning several positivist and structuralist planning principles, interventions and provisions that effectively consider informal development of the city as an 'exception' and a negative alteration of urban development are evident in subsequent development programmes and plans, which sought to intervene through regularization and formalization[23] as the only solutions for reducing people's vulnerability to environmental, economic and social conditions. The policies and interventions that have been implemented come into conflict with a dynamic and rapidly growing city, in which so-called informal settlements and environmental management practices, as well as the presence of hybrid–rural–urban activities and forms, are a widespread and diverse reality that characterizes the modalities of production of space in the city. This therefore requires specific consideration.

A proliferation of plans, ranging across different sectors and scales, have also been designed and partially implemented in the recent years. Those plans, mainly supported by donors and development agencies adopt different approaches; however, they ground on some common assumptions such as legitimacy of legislation, legitimate government, public control of land development, key role of participation in planning system and capacity of government authorities (Berrisford 2014). Legitimacy of legislation assumption falter due to the coexistent of at least two 'laws', the formal and the informal, which lead to an hybrid and dynamic contexts where legal and legitimate are two competing and embedded systems. Moreover, the legitimacy of government is undermined by political tension between national and local government and the control of land development is very difficult where the land marked is driven by local and international investors' interest, poor people looking for a better condition, corruption and a fast growing population. In this environment, it is also exceptional that stakeholders (e.g. individuals or communities) participating in the plan influence the outcome of the plan. In addition, the weak capacity of local government and the lack of financial resources make almost impossible that local council is able to support or align urban management and infrastructure programming with planned expansion and probable future consumption.

[23]A formalization program for economic activities and the use and occupation of land is also present in Tanzania (MKURABITA), which is oriented to defining the areas to formalize, informing and educating citizens and city leaders in order to promote the process, defining regularization schemes that must be approved by governing authorities, and surveying and registering lots for the concession of occupation certificates. This process is not favoured by those who continue to buy, sell, and use land in an informal way (see Chapter 4).

4.4 Which Environmental Transformations: Global Changes and Local Effects in Sub-Saharan Cities

We have seen how the cities of sub-Saharan African are characterized by rapid growth that is unaccompanied by the development of 'urban' infrastructure and dwellings, by serious socio-economic disparities, by limited governability and environmental degradation. These conditions increase sensitivity to environmental transformations, and climate change can worsen environmental, social and economic situations that have already reached a critical level.

The main impacts on sub-Saharan African cities include those deriving from the occurrence of extreme events that are striking the tropical belt of the African continent in an increasingly intense manner (El Nino, storms, severe drought, flooding, etc.), and the environmental stress connected to the availability of water, which is a serious strain on people dependent on natural resources and on the food production system, and therefore food security. In addition, heat waves, air pollution and flooding can heavily condition the health of the urban environment, leading to serious impacts on people's health, which are also linked to the spreading of vectors of disease according to climate variations. Finally, there are numerous impacts on rural–urban connections at the regional (between rural and urban areas) and local level (between urban areas and the peri-urban fringe). Access to resources and the possibility of diversifying income sources is often based on these connections, which can therefore be thrown into crisis as a result of environmental stress with consequent conflicts over resource use and migratory phenomena on various scales.

Dar es Salaam is subject to a variety of environmental changes linked to the combination of its geographical and environmental characteristics and the modalities with which the city has and is being developed.

4.4.1 Environmental Transformations and Climate Change in Dar es Salaam

In Dar es Salaam, the problems caused by climate variability (especially flooding of the coastal zone during the rainy season, see Fig. 4.2) have been at the centre of the urban agenda for some time. Extreme events such as the 2004 Tsunami or El Nino (1992–1993 and 1997–1998) (Shemsanga et al. 2010, cited in URT DEO 2011) have produced a series of damages to coastal infrastructure. In addition to continued stress linked to climate change, including the modification of rainfall patterns and the acidification of the ocean with its destructive effects along the barrier reef, several phenomena of environmental degradation that are already underway as a result of the rapid urban development process (water pollution due to the lack of purification of urban sewage, intense fishing and agricultural activities conducted with improper modalities, multiplication of water extraction wells) are expected to worsen (coastal erosion and salinization of groundwater) (Dodman et al. 2009).

Fig. 4.2 Flooded road in Kimbiji, Temeke Municipality. *Source* Ricci (May 2014)

The essential data on the main climate variations that are relevant for the city are listed as follows:

- Rainfall variability: there has been a decrease from 1200 mm annually in the 1960s to 1000 mm in 2010. Furthermore, increased unpredictability of rainfall has been noted, with temporal shifts in the rain peaks and notable impacts on agriculture and people's livelihoods;
- Temperature increase: the minimum temperature level was below 20 °C between 1979 and 1986, while it rose to in 21 °C 1989 and between 20 and 21 °C in 2000, and since 2000 it has continued to increase (TMA 2011). The average temperature increased by 0.2 °C between 1979 and 2009, while the average minimum temperature between 1979 and 2008 increased by 2 °C.
- Rising sea levels: the global rise revealed by the IPCC (2007) was about 3.1 mm annually during 1993–2003, while that rise decreases to 1.8 mm annually if one considers the longer period of 1961–2003, reflecting an acceleration in recent decades. This is particularly significant for Dar es Salaam and coastal ecosystems (mangroves, coastal wetlands, cliffs, vegetation, tides, etc.) and may, according to some authors, continuously modify the coastline, destroying coastal infrastructure and pushing inhabitants to migrate. Such transformations are already evident in some areas of Dar es Salaam, such as the Kunduchi ward on the northern coast.

There considerable uncertainty regarding climate variations, but the sea level rise is particularly controversial. Indeed, measurements of sea level show a pattern of decrease in Dar es Salaam (Kebede/Nicholls 2010).

Moreover, a study linking urban sprawl to climate change vulnerability in Dar es Salaam, suggest that almost 3 million people (two-thirds of Dar es Salaam's population) are 'in a trajectory of vulnerability as regards their access to water, and they are expected to increase in number'(Macchi/Congedo 2015). Groundwater salinization, due to the combined effects of urban expansion and climate change, seriously affects communities in the coastal plain as they depend heavily on boreholes for accessing to water for domestic and productive purposes (Sappa et al. 2014).

4.4.2 Planned Adaptation and the Role of Local Institutions in Dar es Salaam

The data analysis, as well as several other studies (Dodman et al. 2011; UTR DEO draft 2011), suggests that residents, especially in the peri-urban areas of Dar es Salaam, are aware of the environmental changes underway. In order to face some of those changes and to facilitate environmental protection (e.g. tree planting), residents are also implementing strategies, which are useful to varying degrees in mitigating the impacts of environmental change (Mbonile/Kivelia 2008).

Several similar strategies, though not explicitly connected to adaptation objectives, are being adopted at the institutional level, including guaranteeing the provision of infrastructure and services for the poorest populations. Nevertheless, an 'adaptation deficit' still persists in the city (Dodman et al. 2011).

The Government of Tanzania had already adopted the National Environmental Action Plan in 1994 (UTR 1994), laying the foundations for the subsequent national environmental policy. Tanzania took its first step towards addressing the issue of climate change with the ratification of the UNFCCC in 1996, which was concretized with the drafting of the National Adaptation Programme of Action (NAPA) in 2007. The NAPA recognizes the need to incorporate the impacts of future climate change into public policy.

Many of the adaptation initiatives concern rural areas; nevertheless, in recent years several studies have concentrated on Dar es Salaam (UN-Habitat 2011; Dodman et al. 2011). During the COP 15 in Copenhagen, 2009, the Mayor's Task Force on climate change chose Dar es Salaam as one of the cities where pilot studies would be conducted. The Task Force will concentrate on three aspects: understanding the links between urban poverty and climate change, identifying best practices for reducing urban poverty and vulnerability to climate change and promoting investment programmes in order to support best practices. Other programmes related to climate change that are currently underway include Climate Adaptation through Participatory Research and Local Action, in the Temeke municipality, which is expected to expand into two other municipalities; and the Dar es Salaam Resilience Action Plan (DRAP), which should lead to the development of additional actions related to climate change and the integration of the results of various ongoing studies that are part of the new Master Plan.

Furthermore, the Dar es Salaam city administration has a Disaster Management Unit that coordinates responses to disasters, including those related to climate change, such as fighting cholera epidemics.

Following the publication of the NAPA, a series of studies were initiated in order to assess the city's vulnerability to climate change, and they are still in progress.

An analysis of the plans and the relatively limited strategies related to climate change indicate that it has been difficult to implement them and to find resources. Interviews with employees of local institutions (Kinondoni district and Kunduchi, Bunju and Kawe wards) have revealed, on the one hand, the existence of a large number of plans for environmental conservation and protection, and on the other, that most city planning is still concentrated at the ministerial level, and long-term planning is almost entirely absent. In fact, the interventions in which those employees are involved are generally very specific and linked to responses for resolving contingent problems.

Some plans at the national level that pursue environmental conservation objectives can nevertheless contribute to adaptation to climate change. This is the case as regards the General Plan for the Conservation of Natural Resources, drafted by the Ministry of Natural Resources and Tourism, to be implemented by municipalities in order to preserve the mangroves, control fishing and mariculture practices, and to prevent illegal fishing, unauthorized sand pits and unjustified tree cutting. Although these are not explicitly objectives of adaptation to climate change, many of the planned measures can also reduce the vulnerability of people and ecosystems to environmental change. However, it should be noted that the environmental conservation measures have never been situated in relation to urban development measures, which are principally linked to the development of infrastructure. As a result, institutional behaviour is often contradictory.

An analysis of climate change adaptation policies in Tanzania demonstrates that prevail a depoliticized framing of climate change impacts and a technocratic vision of adaptation. In this context, the numerous pressures on local institutions are likely to intensify and exacerbate existing problems of capacity and transparency that limit their potential to contribute positively to socially equitable adaptation (Smucker et al. 2015).

4.5 A New Environmental Question for an Old Planning Problem

We have seen that the environmental, social and economic effects of environmental change are considerable. The planning policies and strategies underway leave many questions unanswered, which draws attention to several criticalities that are not new to urban studies.

The issue of environment has been at the centre of the contradictions between contemporary urban strategies, as is reflected in Marvin and Hodson's criticisms of securitization approaches (2009). In fact, it would appear that the theories of urban bias and anti-urban bias still constitute the pivotal points around which such strategies revolve, especially for peri-urban areas that continue to be seen as incomplete pieces of the city, or parts of the countryside to reclaim and restore.

Even the issue of informal development of the city is brought to the centre of climate change: that which is informal 'doesn't exist', or exists as a problem to resolve, reclaim and regularize. On the other hand, the informal is also a world of resources and creativity that is capable of self-organization in any situation, and can survive even when institutional contributions are limited.

The planning system in Tanzania has been developed (at least in theory) around a tradition of 'comprehensive planning', first introduced by British colonizers, and later reinforced by the Master Plans developed after independence (MacGregor 1995). In practice, this approach is often substituted and/or supplemented with planning for single interventions, guided by an interpretation that pays little attention to processes and directives in a complicated and fragmented combination of regulations. Given the inefficacy of plans and projects, planners have had to ask themselves why development processes did not follow their long-term forecasts, trapped in the dilemma of having to either 'accept' reality or impose norms.

The attempts to respond to these problems in the disciplinary sphere have reflected several changes, including a growing tendency to place less emphasis on the planning of limitations and control, and a view of planners' input as only one of many inputs necessary in the development process. This has led to new forms of non-technical knowledge and the involvement of community members interested in the definition of a common vision (Healey 1997, 2010). Planning is increasingly understood and practiced as an iterative, participatory and flexible process, despite the fact that in many cases it still proceeds as a process that keeps the analysis of urban and regional transformations separate from the governance processes through which decisions are made.

As we have seen, the environmental conceptualization of the peri-urban interface and the vast interactions between rural and urban that give rise to specific transformation processes of that area have a variety of implications as regards the formulation of planning interventions. The complexity of the phenomena and subjects that characterize the peri-urban interface poses numerous questions for planning, questions of scale, instruments, competency and others.

The development of a new perspective implies a rethinking of intervention policies and strategies: a 'non-urban' interpretation of the peri-urban interface could modify how certain political questions are formulated (the informal, for example) and how certain environmental problems are represented.

How one can arrive at a planning approach that responds to the specificity of the peri-urban areas remains an open question that is widely debated. One contribution that may allow that debate to move forward may come from research into a new interpretation of space that could be the basis for identifying sustainable processes that would need to be preserved in order to develop new ones, which will be essential to improving the living conditions of peri-urban and other inhabitants.

While 'securitization' and urban development represent an indiscriminate solution for any type of urban and peri-urban space, some authors assert that careful analysis of the local community, consideration for participatory models in the public sphere and an understanding of how transformations can be triggered and/or managed is necessary in order to orient planning to the benefit of the most vulnerable groups. A focus on strengthening the organizations of peri-urban residents is also essential in order to render their hopes and needs 'visible' to the various institutions that intervene in this context. Nevertheless, in order to prevent an 'isolationist' approach from impeding the emergence of independent coalitions of diverse social groups, examination of the 'political capacity', both individual and collective, of peri-urban groups and its effects on the management of resources, is necessary (Allen 2006).

Chapter 5
Adaptive Capacity as a Strategic Element for Reducing Vulnerability to Environmental Changes

Abstract This chapter focuses on the role of the autonomous practices for environmental management and adaptation to environmental transformation adopted by people settled in peri-urban areas. This is considered the point of departure for the formulation and implementation of local adaptation measures and action plans in the urban domain. In particular, the chapter identifies the key factors in assessing peri-urban adaptive capacity, and presents the results of the investigation carried out in Dar es Salaam. Modalities of access to and management of resources, perception and observation of environmental transformation, and current or prospective autonomous strategies for confronting it are then discussed. Finally, the relationship between household characteristics and adaptation strategies is outlined in order to identify the specific interdependencies, delicate equilibria, limits, and opportunities that derive from those interdependencies, and to understand how they may influence measures undertaken by institutions to adapt to environmental transformation.

Keywords Vulnerability · Resilience · Adaptive capacity · Autonomous adaptation · Natural resources · Environmental management · Livelihood

5.1 Vulnerability, Resilience, and Adaptive Capacity

According to the conceptualizations developed in the literature on climate change (Smit/Wandel 2006), *vulnerability* is a function of a system's *exposure* and *sensitivity* to dangerous conditions, and its *adaptive capacity* (which some define as 'resilience') with respect to the effects produced by such conditions.

More specifically, *adaptive capacity* refers to the capacity of a system to reorganize itself when confronted with a disturbance or potential damage, to take advantage of opportunities, and to face the consequences of the changes under way[1]

---

[1]A "system's ability to adjust to a disturbance, moderate potential damage, take advantage of opportunities, and cope with the consequences of a transformation that occurs" (Gallopìn 2006: 296).

L. Ricci, *Reinterpreting Sub-Saharan Cities through the Concept of Adaptive Capacity*, SpringerBriefs in Environment, Security, Development and Peace 26, DOI 10.1007/978-3-319-27126-2_5

(Gallopìn 2006: 296). For other authors, this means having the capacity to modify exposure to risks associated with climate change, to absorb and recuperate following the damages that derive from climate impacts, and to take advantage of the new opportunities that arise during the adaptation process (Adger/Vincent 2005: 400).[2]

Exposure and sensitivity are linked to the qualities of a community or group of people, and depend on their modalities of interaction with the environment (the context, broadly speaking) as well as the characteristics of the environmental changes considered. While exposure and sensitivity orient the potential impact of climate change, adaptive capacity (and not intrinsic characteristics) is the element that has most bearing on the reduction of the eventual impacts[3] on social systems of such changes, both in terms of both behaviour and capacity to act.

The various interpretive approaches to adaptive capacity in the urban context are closely related to other concepts, such as resilience, coping ability, adaptability, management capacity, flexibility, robustness, and stability (Smit/Wandel 2006).

From the analysis of these approaches, the concept of adaptive capacity can be traced back to two primary theoretical lines of thought, one linked to the concept of vulnerability and the other linked to the concept of resilience. Before defining how the two concepts influence the definition of adaptive capacity, further discussion of the relationship between resilience and vulnerability is necessary.

5.1.1 Vulnerability and Resilience

The analysis of vulnerability has its roots in studies on evaluation and prevention of risks (hazard-risk), but was later developed in other fields, including geography, poverty and development, food security, and political ecology, all of which have profoundly influenced vulnerability as a concept (Eakin/Luers 2006). In the research that considers vulnerability a key component of risk, where risk is a function of danger, the probability that a dangerous event will occur, and the total damages that the event could cause in a given context (Brooks et al. 2005), emphasis is placed on the characteristics of the biophysical system being analyzed (e.g. the presence of certain land uses, of human settlements, of natural resources, etc.) and on the danger itself (e.g. floods, coastal erosion, hurricanes, forest fires, etc.). More recently, other disciplines have pushed for consideration of the social conditions that render people vulnerable (Adger 2006). More specifically, political

[2]"The capacity to modify exposure to risks associated with climate change, absorb and recover from losses stemming from climate impacts, and exploit new opportunities that arise in the process of adaptation" (Adger/Vincent 2005: 400).

[3]The *Intergovernmental Panel on Climate Change* (IPCC) summarized the determinants of adaptive capacity in the *Third Assessment Report* (TAR) of Working Group II as the following categories: economic resources, technology, information and skills, infrastructure, institutions and equity (Smit et al. 2001). A variety of disciplines have defined and addressed the concept of adaptive capacity, expanding on the contents of the IPCC report.

ecology and geography have focused on 'social vulnerability', emphasizing socio-economic, demographic, cultural, and political characteristics, as well as the role of institutions and governance, in order to define vulnerability in a given context (Adger 1999; Cutter et al. 2003). These approaches emphasize how vulnerability is 'socially differentiated' in an adaptation process, in that it is connected to the security of livelihood strategies, and therefore to availability of and access to resources, which depend on economic and social relations, location, and poverty (Adger 1999: 250).[4] At the moment, there is an increasing body of research on vulnerability, dedicated to dual consideration of the biophysical system and social aspects as elements that render the 'system' vulnerable (Clark et al. 1998; Luers 2005; O'Brien et al. 2004b; Polsky et al. 2007), while there is growing interest in the economic aspect of vulnerability and adaptation (e.g. FP7 ClimateCost project,[5] JRC PESETA II project).[6] Moulded by human actions and influencing both the biophysical and social elements of a system, adaptive capacity is considered crucial for the reduction of vulnerability (Eakin/Luers 2006) since it modifies exposure and sensitivity to changes. Although opposing interpretations with unclear distinctions between exposure, sensitivity, and adaptive capacity continue to be developed (Gallopìn 2006; Fussel 2007), adaptive capacity is generally recognized as a desirable quality, or positive attribute of a system in terms of reducing vulnerability.[7]

Resilience, or the achievement of a desirable state when faced with a change (Folke 2006), has its origins in scientific ecology's attempts to elaborate theoretical and mathematical models (Gallopìn 2006; Janssen/Ostrom 2006). Complexity theory (Holling 1973), systems theory, and agent-based community models have also contributed to the development of the concept of resilience.

Although it originates in natural sciences, the resilience perspective tends to increasingly include human contributions to the dynamics of a system, as is demonstrated by the development of the literature on Socio-Ecological Systems (SESs) (Walker et al. 2006). This approach, which recognizes that the human presence in ecosystems is one of the principle causes of change (Folke 2006), maintains that a study of human and environmental systems as well as the interaction between the two is necessary in order to understand the mechanisms involved within and between systems (Janssen/Ostrom 2006). The unit of analysis in resilience research therefore become a complex combined system (SES) of human components (e.g. institutions,

[4]From this perspective, income and consumption level reflect but are not directly correlated to access to resources. The limits of this relationship depend on what Sen (1984) defines as entitlement (right, attribution): "the set of commodity bundles that a person can command in a society using the totality of rights and opportunities that he or she faces, and which are in fact bound by legality or custom. In other words, opportunities to avoid poverty (such as by raising income) are often constrained by rights to buy or sell resources".

[5]For more information see: http://www.climatecost.cc/.

[6]For more information see: http://peseta.jrc.ec.europa.eu/.

[7]We will see that the literature on resilience also represents adaptive capacity as a desirable property of a system, but with a slightly different meaning.

infrastructure, culture, etc.) and environmental components (e.g. geological, climatological, biological, etc.; Turner et al. 2003; Gallopìn 2006).

Recognizing the limits of incremental adaptation that would only strength existing measures, strategies, and capacities to limit impacts (Kates et al. 2012), the resilience framework assumes transformation as a core concern. Transformation is considered an adaptive change to ensure persistence of existing capacities and functions, including "the interplay of persistence, adaptability, and transformability" (Folke et al. 2010: 25; Field et al. 2014).

In contrast to *transformational adaptation*, the notion of *transitional adaptation* focuses on social change, such as the pursuit of distributive and procedural justice through reform of governance systems. Transitional adaptation raises the question of 'the potential for adaptation' opening existing governance regimes up to reform (Pelling 2011).

If, on the one hand, resilience research continues to encounter the problem of having to give greater consideration to the social aspects of SESs (Adger 2006), on the other, adaptive capacity has started to receive more attention in the literature (Carpenter/Brock 2008; Pahl-Wostl 2009), which describes it as the capacity of actors in a system to manage and influence resilience (Walker et al. 2004), facilitating the interaction between anthropic and environmental components.

Three different tendencies emerge from this continuously evolving panorama of conceptualizations: several authors define resilience as the flip side, or the antonym, of vulnerability (Folke et al. 2002); others conceive resilience and vulnerability "as two overlapping inherent properties of local places [that] together interact with natural hazards and coping responses to produce or mitigate accumulative disaster impacts" (Cutter et al. 2008 cited in Romero-Lankao/Qin 2001: 4); others still identify them as two complementary approaches to the evaluation of adaptive capacity (Engle 2011; Brooks et al. 2005; Fig. 5.1).

In particular, the latter interpretation sees adaptive capacity as the component of vulnerability that can modulate the exposure and sensitivity of a given system (Yohe/Tol 2002; Adger et al. 2007), and as the capacity that renders a system most resilient and therefore more able than others to evolve towards a 'desirable' state. In other words, it is able to modulate between maintaining the status quo and transforming the Socio-Ecological System into a new state, according to which of the two would be more "desirable"[8] (Engle 2011: 5). This means that those who interpret resilience and vulnerability as complementary measure adaptive capacity in terms of the potential of a system to evolve towards a more desirable state.[9]

[8]"Desirability" is a social construct (Robards et al. 2011: 253). According to how it is negotiated within a system, adaptive capacity is configured as the maintenance or transformation of the status quo.

[9]There are nevertheless authors who view adaptive capacity as the antithesis of resilience: while the former is identified with change, the latter is associated with entrenchment (Smithers/Smit 1998, in Schoon 2005). This interpretation likens resilience to the capacity of a system to return to its pre-existing state after a given disruption, and does not contemplate the possibility of achieving a state of equilibrium that is different from the original one.

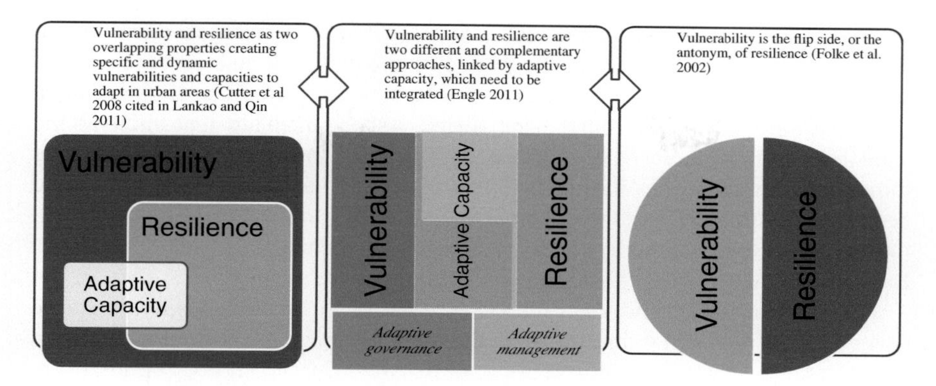

Fig. 5.1 Approaches to the relationship between vulnerability, resilience, and adaptive capacity. *Source* The author

The approach adopted in the present research is centred on people, though it also recognizes the complexity and the close interaction between natural and anthropic systems and between human and environmental components. Resilience and vulnerability are not, therefore, intended as properties of the 'urban' system, as this abstraction would impede consideration of certain social justice aspects,[10] and would not allow recognition of the agency of individuals (Romero-Lankao/Qin 2011: 5). This research is focused on the social vulnerability of individuals and groups as a function of their short- or long-term relationships with the environment and natural resources. Therefore, the study also takes into consideration the sustainability of economic, environmental, and social processes and the reproducibility of natural cycles.

5.1.2 Adaptive Capacity and Urban Vulnerability

Research on vulnerability in the urban context (urban vulnerability) is characterized by tension between the need to represent the differences between and within urban areas, which arise from the specific nature, dimensions, and factors of the context, and the desire to identify the determinants and attributes of adaptive capacity and resilience on an urban scale.[11]

---

[10]For example, some of those who emphasize the link between adaptation and development criticize the vulnerability approach for failing to address the structural causes of vulnerability (Sanchez-Rodriguez 2009).

[11]"A set of concepts and tools that cut across knowledge areas is needed to improve the understanding of how urban vulnerability is characterized and determined by issues such as thresholds, tipping points, second and third order impacts, and responses" (Romero-Lankao/Qin 2011: 1).

Studies of urban vulnerability often tend to define it in negative terms, as the possibility of being damaged, or the degree of susceptibility and the capacity of a system (city, population, infrastructure) to face the negative effects of one or more risks and stresses. On the other hand, many authors maintain that vulnerability cannot be defined exclusively in terms of risk, nor can it be represented strictly through the intrinsic properties of the system placed under stress. Rather, it should be seen as an interaction of these factors that includes various dimensions: potential impacts, exposure, sensitivity, capacity to adapt and actual response (Adger 2006; Eakin/Luers 2006; Romero-Lankao/Qin 2011).

For the most part, the principle lines of research on urban vulnerability to climate change reflect those on vulnerability in the more general context of environmental change (Adger 2006; Eakin/Luers 2006; O'Brien 2007):

- urban vulnerability as impact (analysis of natural risks);
- contextual urban vulnerability (political economy or ecology);
- vulnerability in relation to the capacity to respond (ecological resilience).

In the literature on natural risks, urban vulnerability to climate change is conceptualized as impact, or the result of exposure to dangerous climate events and the sensitivity of the urban context (understood as infrastructure, population, and activity). Within this sphere, there are two major lines of research.

The first analyzes how the variation of a climate parameter or a combination of parameters (e.g. temperature, air pollution, precipitation) is connected to specific impacts (e.g. increased mortality), and explores how the characteristics of a population (e.g. age, gender, socio-economic status) influence the overall impact under consideration, identifying those characteristics as risk factors (Ruddell et al. 2010 in Romero-Lankao/Qin 2011: 2–5). Other studies examine the geographic characterization of urban settlements (e.g. metres above sea level, slope, water scarcity) that render residents of urban areas vulnerable to the impacts of climate change (McGranahan et al. 2007).

The second line of research, defined as top-down impact evaluation, applies climate change scenarios and adapts them to the urban scale in order to model how certain parameters (e.g. temperature increase and sea level rise) will evolve in the future. In some cases, adaptation options are also explored.

Urban vulnerability as impact addresses questions such as exposure or sensitivity to variations in risk factors. By exploring the nature of risks (intensity, frequency and duration), impact assessment is also useful in addressing political questions, as it leads to investigation of which impacts could be avoided through adequate political response (Romero-Lankao/Qin 2011: 2–5).

Nevertheless, the application of this approach in the urban context is criticized for several reasons. First, because it omits analysis of the structural causes of vulnerability, such as the reasons for which specific urban centres, populations and sectors within cities are affected differently by changes. This means it does not consider if and when local stakeholders and inhabitants are receptive to certain planned adaptation interventions, motivated to make the necessary changes, or whether they possess the necessary capacities, knowledge, means, and resources that would allow them to adapt. Finally, it does not take into consideration how their potential adaptation choices might be limited by the social, economic, political, and environmental circumstances in which they live and act (Romero-Lankao/Qin 2011). In other words, because the approach of vulnerability as impact omits the aspect of agency, institutions, informal organizations, individuals, and households are represented as passive receptors of stress and risk, and not as subjects that are actively responding to climate change and environmental transformation.

In order to overcome some of these shortcomings, an approach has been developed that views urban vulnerability as 'contextual' and that is oriented towards understanding the structural reasons (environmental, social, economic, and political) for which certain cities and populations are more or less vulnerable than others (Adger 2006; Eakin/Luers 2006).[12] Within this approach, which is an evolution of the Livelihoods Approach, the theories of political economy and the approaches of ecology and political ecology, there are several differences that are attributable to the assumptions of various analyses and interventions. A focus on livelihoods, on the need to explore and reaffirm the construction of the subsistence assets of individuals, households, and communities, is generally seen as a fundamental mechanism that allows individuals and households to face the risks that they constantly encounter (Enlge 2011). Nevertheless, some authors affirm that the Livelihood Approach lacks adequate recognition of the role of the State in the development of people's adaptive capacity, through for example the promotion of economic growth and poverty reduction. These researchers have demonstrated that interventions that concentrate on the role of institutions, with approaches based on rights and distributive justice, are indispensable to the construction and maintenance of structural determinants of adaptive capacity (Satterthwaite et al. 2007; Parnell et al. 2007) (Fig. 5.2).

The key is therefore to understand what determinants of adaptive capacity planning should act upon in order to reduce social vulnerability in the urban (and peri-urban) sphere. Much of the literature identifies such determinants as the presence or absence of infrastructure and services, soil and housing quality, and public emergency systems. Moreover, the debate on African cities and low- and medium-income countries, which are most affected by climate change, has underlined the existence of multi-level determinants of urban vulnerability that are

[12]From the perspective of political economy and geography, Adger (2003) maintains that a community's capacity to adapt is limited by its capacity to act collectively, which is heavily influenced by social capital, trust relationships, and social organization (Adger 2003; Pelling/High 2005). In order for the planned (or anticipated) adaptation actions to be successful, preparation of a series of basic factors is necessary, such as the presence of effective economic structures (Engle 2011: 3).

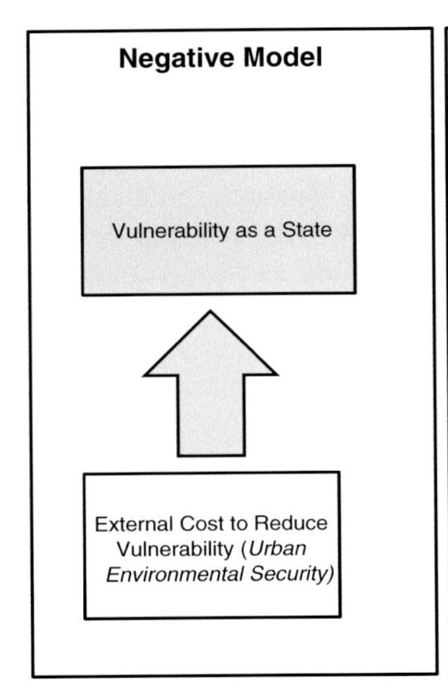

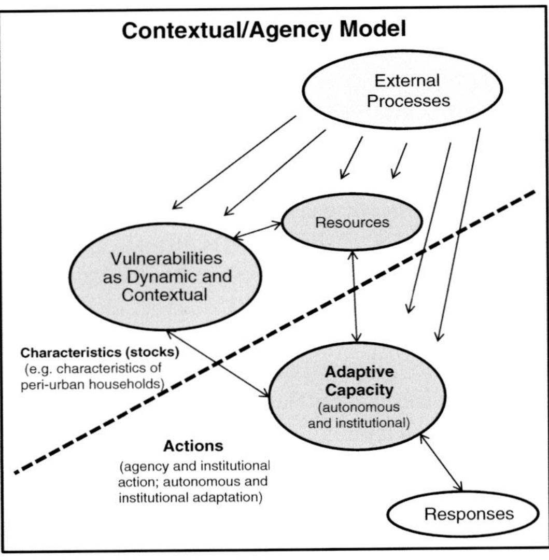

Fig. 5.2 Approaches to vulnerability in the urban sphere. *Source* Adapted from Maguire/Cartwright (2008)

principally connected to the failures of governance and development[13] (Yohe/Tol 2002; Ivy et al. 2004; Brooks et al. 20005; Haddad 2005; Eakin/Lemos 2006; Agrawal 2008; Brown et al. 2010; Engle/Lemos 2010; Gupta et al. 2010 in Engle 2011: 3).

In light of the considerations on the 'functioning' of African cities provided in the second and third chapters, one wonders whether it is only these key components that are to be analyzed, or if there are other aspects to which the critical approaches of 'Asymmetrical Ignorance' and 'People as Infrastructure' should draw more attention.

To return to Roy's comments (2005: 149) as regards the 'state of exception' (Sect. 2.2.2), when considering the role that the State and institutions play in the development of adaptive capacity, one must first and foremost recognize their relationship with (and not simply the existence of) informal practices and social networks, or with 'platforms of action' (Simone 2004), and the immaterial processes by which they are characterized. Autonomous adaptation practices and strategies are not subject to, but rather are the *product of* institutional regulations. In this sense, they are legitimate, even if they are not regulated by formal (legal) norms, and

[13]From this perspective, an important and increasingly vast body of literature addresses how institutions, governance and management play a fundamental role in determining the capacity of a system to adapt to climate change (Engle 2011).

are therefore an expression of what is defined as sovereignty. This perspective allows one to call into question the boundary between autonomous adaptive capacity, or spontaneous informal adaptation practices, and institutional, planned or formal adaptation. This boundary, if it exists, is so blurred that it leads one to rethink adaptive capacity in different terms.

On the one hand, criticism of approaches to urban vulnerability to climate change has prompted research into new perspectives that are able to integrate the diverse dimensions of vulnerability and adaptation, and thus support the definition and actuation of more effective responses to environmental transformation. On the other hand, post-colonial and human agency theories provide an alternative perspective from which to understand the types of strategies that individuals, communities, and institutions need in order to face and adapt to environmental change in urban and peri-urban areas.

5.1.3 Linking Peri-urban Dynamics and Adaptive Capacity

The rural–urban interface is particularly challenging when addressing adaptive capacity concerns. While the environmental change in African cities (and in Least Developed Countries in general) is broadly studied, these issues have received little attention in peri-urban areas, despite the fact that they constitute the most at-risk part of African cities.

Peri-urban areas should be treated as integral elements of urban systems in spatial, social, economic, functional, and planning terms, because they and their environments are integral to the growth and operation of growing cities (Simon 2008). Those same peri-urban areas include features that are relevant to rural development and livelihood policies, and are also the place the place with the most potential for positive change, due to the many forces that come together in such spaces (Erling 2007, cited in Simon 2008). While explaining the nature of this potential, Erling gives the example of multifunctional urban agriculture as a source of creative and ingenious new approaches to producing food amid competition for land use. Others (Davila 2002; Allen 2006) argue that there are other opportunities for positive change that could be useful for adaptation to environmental change, such as livelihood diversification, access to services, reuse of waste in agriculture, waste recycling for re-sale in the city (though this might be associated with health problems and pollution), and greater access to information and decision-making.

According to the definitions of vulnerability reported in Sect. 5.1, while exposure and sensitivity orient the potential impact of climate change, adaptive capacity can be a major influence on the eventual impacts of those changes. Adaptive capacity is therefore an obvious focus for adaptation planning because it is the component of vulnerability most amenable to influencing social systems coping with climate changes (Marshall et al. 2010).

5.1.4 Assessing Adaptive Capacity

We have seen that conceptions of vulnerability and resilience are closely connected to those of adaptive capacity, and therefore to the various approaches for evaluating it.

As mentioned in Chap. 1, the research process has been constructed on the hypothesis that adaptive capacity to climate change in peri-urban areas depends on four main factors:

(1) Typology and extent of environmental impacts of climate change at the local level;
(2) Rural–urban dynamics and relationships land use to the urban fabric;
(3) Autonomous local capacity to cope with the consequences of climate change;
(4) Institutional capacity in environmental and urban management and planning.

The first point is linked to the fact that transformations of differing typologies and intensities are linked to different capacities to cope with them. This means that it is not possible to establish a generic adaptive capacity to environmental change, but only one that is specific to a certain type of transformation. Points 2 and 3 are linked to the fact that in contexts similar to peri-urban areas of sub-Saharan cities (but not exclusively), where there is a heavy reliance on natural resources, adaptive capacity as expressed in terms of livelihood strategies depends closely on rights and mechanisms of access to resources (entitlement theory). The integration of this concept with other aspects of political ecology, which includes social and power relations in order to explain the decision-making process, allows for the elaboration of a theory that is constructed on the multi-dimensional differences of society (based on economic level, gender, age, identity, etc.). From this perspective, people do not simply draw from their own assets and resources, rather they possess sophisticated competencies for managing them, coping with adversity, adapting with flexibility, and taking advantage of new opportunities on diverse temporal scales. The dynamism and diversity in their capacity to mediate allows households to adopt diverse adaptation options for their own livelihood strategies according to environmental and climate shocks and variability over different periods.

The fourth factor is linked to the role that institutions play in the development and implementation of strategies that support adaptation, where such strategies are understood as specific and separate initiatives, but are also part of the mainstreaming of adaptation efforts in the planning and management processes already underway. If institutions are conceived of as the rules and models of behaviour that give form to social interaction, and organizations are understood as groups of individuals connected by a common goal, both institutions and organizations can facilitate collective action and allow individuals to transcend the limits of isolated action.

An understanding of the support, functioning and interaction of the mechanisms of institutions and of social organization, specifically those dependent on natural resources, is therefore of fundamental importance for adaptation. 'Autonomous' and 'institutional' capacities related to factors 3 and 4 are therefore not only connected, but also interdependent and closely intertwined through reciprocal reproductive mechanisms.

The investigation undertaken in Dar es Salaam addressed all four points of the research hypothesis from the perspective of the households interviewed. The extent and typology of the environmental changes referred to are those observed and experienced by the households interviewed (with confirmation in the literature); in the same manner, the urban–rural interactions are those revealed through the administration of questionnaires (and field observation). In this sense, modalities of access to and management of resources are part of adaptation strategies, but are also a link between the formal and the informal, an expression of how the institutional dimension is an integral part of the informal mechanisms of the management of space and vice versa.

The changes observed, the urban-rural interaction and structure, and the modalities of access to and management of resources are defined as elements of the vulnerability context (independent variables) that certainly influence autonomous adaptation practices (dependant variables). The relations between dependant and independent variables demonstrate the critical points upon which institutions can act in order to catalyze positive adaptation processes and actions, and to inhibit those that are detrimental. In other words, to reduce social vulnerability and orient urban development processes towards 'desirable' horizons.

5.2 Investigating Adaptive Capacity in Dar es Salaam: The Search for Key Factors

In order to understand at what point planning processes can intervene, and how autonomous adaptation and environmental management practices interact with the characteristics of households and peri-urban areas, the modalities of access to and management of resources used by the households interviewed were first analyzed, followed by their autonomous adaptation strategies, thus situating those strategies in relation to each other and to the urban–rural interactions described in Chap. 2. This type of analysis is considered crucial for identifying possible adaptation options and/or planning and environmental management measures that do not compromise the livelihood strategies of people living in peri-urban areas. The connections and intersections between modalities of informal environmental management and formal regulation of the use of space and land will now be analyzed.

As already mentioned, the questionnaire was subdivided into four sections:

1. Urban–rural interaction
2. Access to resources and environmental services (land, water, energy, etc.)
3. Management of resources (or environment) (water, waste, land, etc.)
4. Climate change: environmental changes and autonomous adaptation strategies.

The results of the first section were analyzed in Chap. 3. The present chapter will examine the results from Sects. 2, 3 and 4.

The responses from the second and third sections highlighted the modalities of access to natural resources (water, land, energy), the presence or lack of services related thereto (water and electricity grids, collection of solid and liquid waste, etc.) and inhabitants' autonomous modalities of resources management (e.g. water storage, waste management). The various modalities are analyzed in terms of pressure on the environment, social connection and health risks. When comparing the results of analysis of the wards that have urban–rural characteristics (Kawe, Kunduchi, and Bunju) with the results from the 'urban' control ward (Msasani) (see Fig. 2.1), the following inquiries were made:

(1) Whether there were disparities in access to resources between the peri-urban and urban areas, and to what extent the peri-urban areas were disadvantaged as compared with the urban one;
(2) Whether there were disparities in the management of resources and the presence of services and infrastructure;
(3) What the consequences were of such disparities in terms of the impact on human health, and as regards households' livelihood strategies in relation to the insecurity of services and the modalities of access to and management of natural resources;
(4) Whether and what environmental pressures were generated by inadequate resources management, and what consequent risk there was of amplifying the impacts of environmental and climate change.

The answers to the fourth line of inquiry allowed for a determination as to whether the inhabitants of peri-urban areas were more affected by environmental changes than those of more urbanized areas, or whether they tended to over- or underestimate the changes under way because they were in particularly vulnerable circumstances. Therefore, by specifically addressing the autonomous adaptation strategies adopted to cope with the environmental changes observed, the questionnaires allowed for the formulation of a reply to the following questions:

(1) Is it true that the inhabitants of peri-urban areas have a certain capacity to implement 'autonomous' adaptation strategies'?
(2) Is it true that the most widespread of such strategies is that of diversifying income sources?
(3) Do those inhabitants tend to respond to environmental change by adopting lifestyles that are more urban, or rather by modifying their activities and environmental management modalities, possibly moving just to be able to maintain their peri-urban lifestyle?
(4) What dependence exists between an individual's primary economic activity and the adaptation strategies they adopt?
(5) Is it true that, when faced with environmental stress, those who are engaged in agricultural activities tend to move to another area that allows them to continue the same activities?

(6) Is it true that those who work in the urban sphere do not adopt adaptation strategies that involve any modification of the kind of activity they practice in their living environment?

5.2.1 *Resource Access and Management*

The sections of the questionnaire regarding resource access and management aimed to investigate the modalities of accessing resources such as water, land, and energy, and the presence or absence of related facilities, including solid and liquid waste collection. Those aspects are crucial to identify obstacles and opportunities in adaptation to climate change in peri-urban areas. Access to land, water, shorelines, sea and raw materials, energy, and services such as waste management is often an 'informal' process, based on social networks and direct relations with the environment. Therefore, resource management modalities have a considerable effect on the vulnerability of residents and constitute determining factors in adaptive capacity. The identification of resource use and the management regime is essential to understanding households' capacity to interact with the environment, substitute or integrate facilities with their practices, and cope with a lack water supply, waste management or other services. The high level of diversification in resource management and in modalities of accessing resources, especially water, is crucial in the surveyed areas and it is likely to be a major obstacle or opportunity for adaptation.

5.2.1.1 The Question of Land Tenure

One aspect of access to resources that is widely debated, and linked to the production of income and/or food, is the possibility of having the legal or legitimately recognized right to occupy or use land. The study therefore investigated whether it was possible to have such rights recognized in peri-urban areas, and whether that possibility was more or less limited as compared with urbanized areas like the control ward.

In the control ward (Msasani), the size of the plots available to the households interviewed, understood as the living space and the area around it that could eventually be used for other activities, was in almost every instance less than a hectare, in some cases just a few square metres or simply unmeasured. On the other hand, in the three peri-urban wards under investigation, the size of the plots was greater on average (less than or equal to one hectare for more than a third of the interviewees, and between 1 and 6 ha for the others).

This aspect is tightly linked to the complex question of the land tenure system, which seriously conditions the use of land and the modalities of resource management, a question that would require a detailed investigation that is beyond the scope of the present study. Nevertheless, in order to provide an overview of the

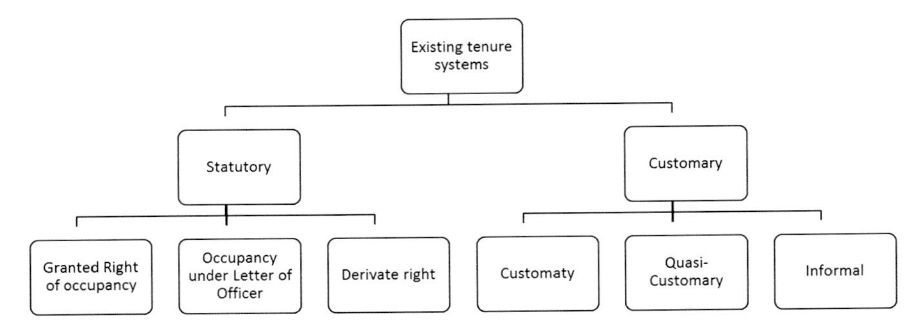

Fig. 5.3 Tanzanian land use regime. *Source* Sheyua (2009, see Annex III)

issue, the elements that are necessary to an understanding of the land tenure regime in Tanzania and Dar es Salaam are introduced here. As previously mentioned, all land in Tanzania is public and formally owned by the President. Private land ownership therefore does not exist, but individual rights to occupy and use land are recognized. There are two main typologies of title that recognize these land use rights: leasehold, a rental title that can be short-term (2 years) or long-term (33, 66 or 99 years), and customary tenure, a title belonging to a tribe settled in a given territory that refers principally to rural areas and is recognized but not guaranteed[14] (see Fig. 5.3).

Of the people interviewed, more than half were not in possession of a land title deed. The few people who asserted that they had a leasehold or other title (rental) therefore live in areas that have undergone a process of parcelling and surveying where taxes are paid in exchange for the provision of electricity and water, if such services are present. Only a few people responded that they were in possession of customary tenure, which derived from their tribe's traditional right of occupation and use. The situation is not much different in the more urbanized control area.

Among the households interviewed, leasehold title is possessed by people who settled in an area more recently (within the last ten years), while customary tenure (or the lack of any land title) is common among those who have been settled for a longer period (more than ten years). This demonstrates that the policies adopted in Dar es Salaam in the last ten years have pushed for regulations oriented to the privatization of land and the abolition of customary tenure in the urban region of the city. Indeed, customary tenure is increasingly uncommon, and although it is recognized in unplanned areas, it is not protected when other land use interests arise (Kironde 2006).

[14]Quasi-customary tenure was also introduced at the end of the 1990s, and represents a decrease in the influence that the clan/community has in land transfer. With this kind of title, local leaders and adjacent landowners must be consulted in the case of land cessation, thus the individual right holder is less protected (Kombe 1995; Kironde 2005).

Though more than half of all households interviewed responded that they were not in possession of any kind of title that certified their right to use the land on which they lived, nearly all of the interviewees had purchased land—usually from another private party and without any kind of formally recognized contract of sale—while only a few households occupied their land without having purchased it, or used it with the permission of the presumed 'owner'. Moreover, only the younger people occupied plots without purchasing them (whether formally or informally). This may be indicative of a progressive stabilization process in which single individuals or households settle informally and later register their occupied plot and request a land title. This process, fairly wide-spread in peri-urban areas (Kironde 2006; Kombe 2005), involves only the households that have an income that is adequate to cover the costs of purchase, and in the case of purchasing land on the formal market, the costs of surveying and land use taxes, which excludes certain segments of the population. The impact of purchase and regulatory costs is also evinced by the fact that those who purchased land (with or without title) generally have a smaller area (up to 1 ha), while the opposite is true for those who have not purchased land.

Nevertheless, even those who are able to cover all the costs often choose not to regularize their land tenure. This suggests that the certainty of being able to occupy and use land (legalization) is not necessarily more guaranteed by the possession of a formal land title deed than it is by recognition of the right to occupy (legitimation) by the community leader and inhabitants of the neighbourhood (defined informal agreements, but in reality tacitly recognized even by the formal system). It is clear that the power relations (economic, social, gender) within regularization and formalization processes (such as the attribution of informal rights), and between the two spheres (formal and informal), play a fundamental role and can therefore generate mechanisms of exclusion and oppression in both cases.

As regards access to network services (electricity, water, etc.), there does not seem to be any direct correlation between possession of land use title and the payment of taxes for the provision of such services. Almost all the interviewees pay taxes on their homes and for the use of land, even when they do not have a recognized title for land use. Similarly, half of the households interviewed pay taxes for electricity, a third for water, and only a few pay for solid waste collection, which in any case is conducted by private parties and not by public services.

5.2.1.2 Access to Water

As regards provision of water for domestic use (but often used for other purposes), the responses from the households interviewed suggest that the main source is street vendors (at a cost of about 1 € per 100 L) or tanker trucks (Figs. 5.4 and 5.5). Nearly a third of the households draw water directly from natural sources such as rivers, wells or holes dug near sewers in humid areas. Only a small percentage of households, residing mainly in the Kunduchi ward, are serviced by the hydro network. Even then, their service is provided intermittently and therefore forces

Fig. 5.4 Water buckets used by street vendors. *Source* Ricci (2010)

Fig. 5.5 Water tanker trucks. *Source* Ricci (2010)

households to resort to alternative sources. Furthermore, in some cases the presence of the hydro network allows for indirect types of access for households that are not directly serviced, but are able to hook up to their neighbours water supply, either in exchange for payment or for free. As regards the urban control area, all households were connected to the hydro network; however, the uncertainty and impermanence of supply forces them to diversify their sources by digging wells, harvesting rain water (Fig. 5.6), or acquiring water from street vendors, as already mentioned.

The difference between the more urbanized area of Msasani and the peri-urban areas is that in the latter there is a greater possibility of diversifying water sources since there is generally a greater availability of natural sources. Moreover, if one were to opt for digging a well, in urban areas there is a greater risk of contamination due to the higher number of pit latrines and other sources of pollution as compared with peri-urban areas.

Fig. 5.6 Rainwater harvesting in Makongo, Kinondoni municipality. *Source* Ricci (2013)

The water obtained or collected via the aforementioned modalities is managed using a variety of methods, depending on need (for exclusive domestic use or, as in the majority of cases, for a mix of agricultural and domestic uses). It is predominantly stored in 20 L plastic containers (buckets or cans) that are easy to transport and can also be used for water collection. Some people integrate or substitute this storage modality with the use of larger plastic tanks (from 100 to 5000 L) or cement tanks, either raised or underground, that have an even greater capacity. Those that have installed cement tanks have mostly been settled for more than ten years. The choice of more capacious and non-removable containers is therefore linked to stable settlement and the desire to improve one's living conditions.

Fig. 5.7 Plastic collection in Kawe. *Source* Ricci (2010)

Fig. 5.8 Compost of cow manure in Mtongani, Kunduchi ward. *Source* Ricci (2010)

5.2.1.3 Waste Management

In peri-urban areas there is no proper garbage collection service. Only a small portion of the interviewees had their garbage collected by private collectors. On the other hand, in the urbanized control area this method is common. Peri-urban areas therefore differ from urbanized ones as regards the possibility of individually managing and making the most of one's own garbage, though this can occur in conditions that generate health and environmental risks. However, even the collection services offered by private collectors generate health and environmental risks because the final location of the refuse is unmonitored; in other words, it is abandoned in areas that are just outside the city and in sewers.

The investigation also found that all households in the peri-urban area manage their refuse autonomously, and that even those who use a collection system only do so for a portion of the refuse they produce. In particular, as regards the diverse modalities of individual management, a majority of the households interviewed burn their own refuse, while a small percentage bury it or abandon it in free areas or sewers. About half the households interviewed draw value from their garbage by recovering certain materials like plastic or metal (Fig. 5.7) (in order to deliver them to an organization that can treat them or reuse them autonomously), by composting, and by recycling organic refuse as fertilizer for agricultural purposes (Fig. 5.8).

The undertaking of such activities, and the recovery of some materials, means that solid refuse is inconsistent. Although there are generally more than six people in a family unit, almost all households in the peri-urban area produce only a small bag of garbage each day, which may be indicative of the fact that having more space makes it easier for them to burn or bury their refuse.

5.2.1.4 Management of Wastewater

Wastewater is also managed autonomously and there are no systems for collecting and disposing of it. In general, large cavities dug in the ground are used until saturated (pit latrines), and only in a few cases is a sceptic tank used. The same modalities are used in the urban area of Msasani, but there the contents of pit latrines and the sceptic tanks are more frequently directed to sewers or artificial canals near the streets.

It should be underlined that refuse management is closely connected to precipitation management. Abundant rainfall, which occurs twice per year, represents the possibility of channelling the contents of pit latrines, as well as other types of solid waste, towards sewers and therefore towards the sea. Obviously, these activities entail health and environmental risks, which are analogous both in the urban and the peri-urban sphere. However, the dumping of blackwater into open-air canals, practiced in the urban areas that have a surface drainage system, represents a danger to health that, due to dilution and to the environment's heightened absorption capacity, seems lower in peri-urban areas. The higher population density and production of refuse in urban areas is not matched by a more organized system of waste collection and treatment, resulting in negative impacts on water and soil quality.

5.2.1.5 Access to Energy

Diversification also occurs as regards energy source. In the urban area, a greater number of households have access to electricity grids as compared with peri-urban areas, where less than half of the interviewees had electricity at home. As in the case of the hydro network, the provision of this service is often intermittent and subject to variations and malfunctions. Thus, when electricity is available it is used for domestic lighting (but often substituted with kerosene), while carbon constitutes the principal source of energy for cooking, and is sometimes used in combination with gas.

The *most significant characteristics of peri-urban households* that emerged from the questionnaire analysis are summarized below, specifically their type of economic activity, the different options for accessing and managing resources, the presence of services and infrastructure, as well as the relation between these options and the type of location. The modalities of environmental management and access to resources have been defined as: autonomous (indicated with (A)) (i.e. not guaranteed by institutional services—some would refer to these as informal); those put in place by institutions (indicated with (I)); and those in which the two modalities coexist (A/I). As regards livelihood activities, whether the economy in which they occur is formal or informal has not been indicated, since for each typology both are possible and co-present.

The *livelihood or income production activities* are:

- Exclusively agriculture or other rural activities that depend on natural resources
- Agriculture and other rural activities as primary but not exclusive livelihood activity
- Exclusively urban employment (not dependent on natural resources)
- Urban employment (not dependant on natural resources) as livelihood activity combined with secondary activity
- Urban and rural employment that contribute relatively equally to livelihood.

The modalities of *access to water* (possibility of hydro provisioning for predominantly domestic use) are:

- Exclusively street vendors or tanker trucks (A)
- Street vendors or tanker trucks and pipelines (A)
- Exclusively hydro pipelines (I)
- Exclusively rivers or other natural sources (A)
- Combination of more than two modalities (A).

The modalities of *accessing land* (possibility of a recognized right to occupy and/or use for residential or productive purposes) are:

- Area with title deed (leasehold or customary) (I)
- Area without title deed (A)
- Area with title deed that is rented to/purchased by people who are not the holders of land use rights (A/I)
- Rental/purchase of area not surveyed by third parties (A).

The modalities of *waste management* (solid and liquid) are:

- Collection and management service (public (I) or private (A)) of solid waste
- Autonomous collection and management of solid waste (burying, burning, abandoning, partial reuse) (A)

- Combination of autonomous and serviced collection and management (A/I).

The modalities of *accessing other services and connected infrastructure* (possibility of accessing water and energy distribution, and transportation) are:

- Electricity grid (I)
- Electricity grid combined with other sources (A/I)
- Exclusively other combined sources of energy (charcoal, kerosene, gas) (A)
- Private transport services (A/I).

Environmental location:

- Coastal
- Humid zone
- Plateau
- Lagoon (Mangroves)
- Near a body of water.

Urban Location:

- Near a transportation hub
- Near principal urban services (university, hospital, etc.)
- Near large markets (Kariakoo, Mwenge, Tegeta).

5.2.2 Autonomous Adaptation Strategies and Practices

It is widely recognized that the residents of sub-Saharan cities have adopted multiple adaptation strategies and environmental management practices to cope with environmental threats, however these activities are still neglected in vulnerability assessment and adaptation planning.

The information on autonomous local strategies for adapting to environmental changes, as established by COP 7 (Decision 28/CP.7), must be considered when identifying priority adaptation actions. The following paragraph aims to provide a better understanding of the environmental changes observed by residents of peri-urban areas, their perception of the causes of these changes, and the strategies implemented to address them in both the short- and medium-term.

The strategies and practices of adaptation to environmental change adopted autonomously by households in peri-urban areas depend directly on the changes observed, on the typology and degree of those changes, and the causes to which they are attributed. Such strategies and practices also vary according to the available options for access to and management of the environment and resources, and according to the type of urban-rural relationship that influences a family's living space.

The different adaptation strategies are considered not only in response to exasperation of environmental problems, such as decrease in water availability and

Fig. 5.9 Mbezi River in Makongo, Kinondoni municipality. *Source* Ricci (April 2012)

Fig. 5.10 Shallow pit for water harvesting in Madale, Kunduchi ward. *Source* Ricci (2010)

deterioration of environmental conditions, but also in the event of higher population pressure and new urban developments, which would interfere with ordinary practices and activities (e.g. agriculture, livestock keeping, etc.).

5.2.2.1 The Changes Observed

A majority of the interviewees had recently observed changes in the availability of water, in the fertility and aridity of the soil, in humidity of the air, and in rainfall.

Their responses to the questionnaire suggest that water availability has decreased considerably in recent years. Rivers with a continuous flow (e.g. Mbezi river) from which water could be drawn year round have become seasonal, filling only during the two rainy periods (Fig. 5.9). In areas where water for domestic use is obtained from shallow pits dug in humid zones or near riverbanks (Fig. 5.10), reduction in the amount of water has necessitated the digging of cavities that are deeper and in new locations.

In recent years soil fertility has also decreased, aridity has increased, and air humidity has undergone similar changes.

Moreover, changes in rainfall have been observed by nearly all the interviewees, both in terms of quantity (decrease) and patterns. Many people noted that rainfall no longer follows normal seasonal patterns, with changes in timing, location, intensity and frequency that have forced people to organize their activities differently over the course of the year, for example by planning agricultural rotations differently.

Further changes, such as the receding shoreline and the decrease in biodiversity, have been observed by a small number of people, in part because such changes are closely linked to the environmental context and the morphology of the terrain (coastal areas, plateaus, etc.). In a few cases the changes observed have been considerable, meaning there has been a substantial deviation from normal environmental dynamics. The residents of a coastal area, for example, asserted that the coastline had receded considerably (about 100 m in the last 30 years according to one elderly fisherman from Mtongani, in the Kunduchi ward), changing the morphology of the settlement and the employment modalities of fishermen.

It is important to note that these and other environmental changes are the effect of a combination of factors that are not exclusively linked to global warming, but also to anthropic pressures and, in some cases, specific urban or environmental policies.

Almost all respondents noted changes in water availability, soil fertility, soil aridity, air humidity, and rain patterns as well as other changes such as sea level rise, biodiversity, extreme events, and shifting ocean currents (the results are consistent with those of the survey administered to 6000 households under the ACCDAR project). According to the results of the survey, availability of water has been declining significantly and significant changes were observed both in the amount of rain and in normal seasonal rainfall patterns (Ricci et al. 2012a).

The same survey highlighted that observed changes are also linked to households' activities. Households engaged in agriculture and livestock keeping as their main activities observed more changes in air humidity (decreasing), rain patterns (decreasing) and other environmental changes than those engaged in shop/small business activities or working in urban areas. The households dependent on natural resources note more and significant changes while those who are independent of natural resources do not. In general, households observing more environmental changes also have a lower mean income level (Ricci et al. 2012a).

Fig. 5.11 River embankment on Mbezi River, Kunduchi ward. *Source* Ricci (2010)

5.2.2.2 The Strategies Implemented

More than half the people interviewed claimed to have adopted some kind of strategy for coping with the environmental changes under way. Due to a decrease in the availability of water, some people have changed crop (e.g. switching from rice, which requires a large amount of water, to cassava, which needs less), or approach to livestock (e.g. switching to free husbandry). For the same reason, others have decided to change their type of activity (e.g. switching from agriculture to livestock, from fishing to agriculture, or abandoning 'rural' activities in favour of something less dependent on natural resources and environmental conditions, or looking for temporary employment). In some cases new environmental management practices have been undertaken. For example, in rivers with little water, small accumulation pools have been created through the construction of small retention embankments using earth or other materials in order to prevent the little water that still flows from being absorbed by the riverbed (Fig. 5.11). Another strategy includes community organization to build and manage rainwater harvesting ponds (Fig. 5.12) This allows for continued use of water for agricultural or domestic purposes, even during the dry season, but it may have impacts on the river system and on the water cycle.

The changes in activity or habits occur in a higher percentage of those who have larger plots of land (more than one hectare), while those who claimed not to have adopted any changes generally live on plots that are smaller than a hectare. This probably derives from the fact that those who live on larger plots have an increased likelihood of observing or are more exposed to environmental changes.

Fig. 5.12 Rainwater harvesting pond in Madale, Kinondony municipality. *Source* Ricci (2009)

5.2.2.3 Options for the Future

Many people have considered the possibility that environmental conditions will worsen, having observed more rapid and consistent changes in recent years as compared with the past, and they have therefore hypothesized potential strategies for coping with further difficulty.

Such strategies are in part similar to those already adopted, such as changing activity, searching for temporary employment, switching from a livelihood activity reliant on natural resources to a trade or a small business only partially or indirectly dependant on natural resources (e.g. a small restaurant, or selling handicrafts). In the urban control area, adaptation strategies that anticipate changes in livelihood activity or environmental management are less common and less diversified; however, where they are present they are similar to those mentioned in the peri-urban areas under investigation.

In some cases, the possibility of worsening environmental conditions that are not compatible with lifestyle expectations leads individuals to consider moving to another area, or returning to their place of origin. It should be noted that such strategies are conceived of not only in the case of deterioration of critical environmental factors linked to climate, but also in the event of an excessive urban development of the area that would prevent a continuation of an individual's activities, forcing them to adopt an 'urban' lifestyle. In this case, an important difference arises between the peri-urban areas and the urban control. In the former, there is once again a diversification of strategies that includes the possibility of changing or adapting one's employment activity and environmental management practices in addition to moving to a new location. On the other hand, as a result of the combined effect of urban development and environmental pressures,[15] in the more urbanized area the most common strategy is to move to a peri-urban area where agricultural and livestock practices and the construction of a shelter are possible.

According to the survey administered under ACCDAR project, among respondents, 43 % of respondents have thought of adopting adaptation strategies in the future if environmental conditions worsen. These respondents would choose to start a new business (30 %), move to another area (26 %) or undertake other changes (Ricci et al. 2012a).

Moreover, analyzing the possible connection between adaptation strategies and main sources of income, it has emerged that households practicing rural activities (agriculture, livestock keeping) and other activities that are dependent on natural resource (fishing) prefer to continue with the same changes they are currently

[15]For example, in Msasani Bonde La Mpunga, a densely built-up area where agriculture is still practiced, certain areas are subject to flooding, and in some cases people are forced to fill in the entire ground floor of their dwelling (with earth, sand, etc.) and to live only on the second floor, if it exists. These conditions have been caused by the construction of buildings, not far from the neighbourhood and close to the coastline, that have obstructed natural water drainage and created an accumulation zone at risk of flooding.

making as regards livestock and crops, or to intensify the activity that they are already practicing. On the other hand, those engaged in other activities, such us charcoal making, would prefer to start a new business (Ricci et al. 2012a). In the same study, three household groups were identified: dependent, partially dependent, and not dependent on natural resources. Results shows that those who are dependent on natural resources continue with the same changes they are already undertaking or intensify their activities. Those who are partially dependent on natural resources[16] choose more flexible strategies, such as taking out a loan or getting a temporary job, while households not dependent on natural resources prefer to migrate (Ricci et al. 2012a).

5.2.2.4 The Identified Causes

This last aspect is linked in part to the identification of the causes of the environmental changes under way. More than half of the households interviewed attribute those changes to land use and anthropic local actions (predominantly rapid urban development processes), in combination with or exclusively defined as global warming, which is more or less caused by man.

On the other hand, only a small percentage of the interviewees attributed the changes under way to poor resource management by institutions, or to other factors such as extreme weather events, including *El Niño* in 1997 and the tsunami of 2006, after which changes were observed (e.g. alterations in ocean currents).

Listed below are the *autonomous adaptation strategies* that will be situated in relation to the characteristics of peri-urban households.

Changes to livelihood activities:

- Changes in crop (e.g. from rice to cassava)
- Increase in the surface area cultivated
- Introduction of other income-earning activities
- Improving and intensifying agriculture (e.g. through the use of various agricultural techniques and/or more fertilizer).

Changes to the modality of accessing resources (especially water):

- Digging cavities in humid areas or near the riverbed to extract water
- Multiplication of sources (of water, energy or food)
- Requesting a loan.

[16]It has to be considered that households partially dependent on natural resources are engaged in rural activities as secondary sources of income, and it could be the case that those activities are already part of their adaptation strategy.

Changes in the environment and in the management of resources (e.g. changes in the morphology of the terrain):

- Construction of small embankments to retain water in rivers
- Construction of small channels for water drainage
- Modifications in the structure of dwellings
- Construction of a new dwelling.

Transfer to another area/region:

- To an area with more surface area for agriculture
- To an area with a more fertile agricultural area
- To an area closer to main thoroughfares or transportation hubs (where it is possible to develop a small business)
- Return to place of origin.

Changes to livelihood activity (to something less dependent on natural resources)

- Switch from agriculture to livestock
- Abandon agriculture and/or livestock to undertake another activity that is not dependant on natural resources (e.g. small business, trade)
- Switch from fishing to agriculture
- Search for temporary employment.

5.3 Adaptation and Environmental Management in Dar es Salaam

As has emerged in the above paragraphs, both access to resources and environmental management are undertaken in a predominantly autonomous (informal) way, both in peri-urban areas and in the urban area used as a control for this study, which is consistent with assertions regarding many cities in sub-Saharan Africa (see Chap. 2). Although the urban area has more infrastructure, the provision of resources and environmental management use a system of relations and actions that are defined as informal and are similar to what occurs in peri-urban areas, where infrastructure (water pipelines, electricity, etc.) are virtually non-existent. This similarity is attributable, first and foremost, to the unreliability and lack of efficiency of the services and infrastructure in urban areas, which are unable to guarantee continuous access for inhabitants, who are forced to look for alternative solutions for the provision of water, energy, etc. Second, there is probably a tradition linked both to the physical-environmental characteristics of locations (climate, morphology, vegetation, etc.) and to the cultural factors upon which inhabitants' modalities of action and organization depend.

In light of these considerations, the main difference between urban and peri-urban areas is the higher number of opportunities and options available to the latter in terms of managing and accessing resource, which are becoming ever scarcer, and in terms of reacting to a changing environment.

As regards management and access to resources as well as adaptation options, it is important to note that in both urban and peri-urban areas, none of the interviewees ever referred to the need or the desire to use new or better infrastructure provided by institutions. They only discussed the options that had been thought of, and could be put into practice, on the basis of relationships with the environment and with people. This is probably linked to an awareness that strained financial resources limit local institutions' interventions to emergency situations. It is probably also linked to a habit of relying on a dynamic and consolidated system of organizations and relations that interacts with the 'formal' work of institutions through individuals who belong to other local institutions and are firmly rooted in the territory, such as community leaders or street leaders (*mitaa*).

One cannot assert that peri-urban areas have scarcer or more limited access to resources as compared with more urbanized areas, though this is commonly claimed to be the case. Though the Msasani area (urban) has infrastructure for providing water and electricity, at the same time it is subject to higher demand that is not satisfied by services, which are allocated in an unreliable manner, forcing inhabitants to resort to street vendors and tanker trucks in order to acquire water. Meanwhile, in peri-urban areas people can also use the natural sources available to them (wells, rivers, etc.).

Evaluation of vulnerability to environmental change is directly linked to these elements. There does not appear to be a link between a higher level of urban development (i.e. superior urban infrastructure) and a reduction in the vulnerability that results from limited access to resources. Evaluation of vulnerability to environmental change is directly linked to these considerations.

More options for water provisioning actually constitutes a greater possibility of maintaining one's lifestyle or adapting in the case of environmental stress, and may allow a person to continue practicing activities that are dependent on that same resource (small agricultural activities, etc.). The fact that peri-urban areas that are not equipped with infrastructure are not always more vulnerable and exposed to environmental change is further confirmed by the fact that the majority of the people living in urban areas expect to move to another less 'urbanized' area, a phenomenon that does not occur in peri-urban areas.

This does not mean that the 'platforms of action' that are present in peri-urban areas can substitute or operate more efficiently than urban infrastructure. The negative impacts that they may have on the environment and people must also be considered, and each and every modality of accessing and managing resources is associated with a different system of relations and actions. Nevertheless, in contemporary Dar es Salaam, as in many African cities, these platforms of action are in fact the basis of people's livelihoods, and they cannot be replaced from one moment to the next with the introduction of 'modern' infrastructure without considering how their intangible dimensions of spatial production will be substituted, eliminated or modified.

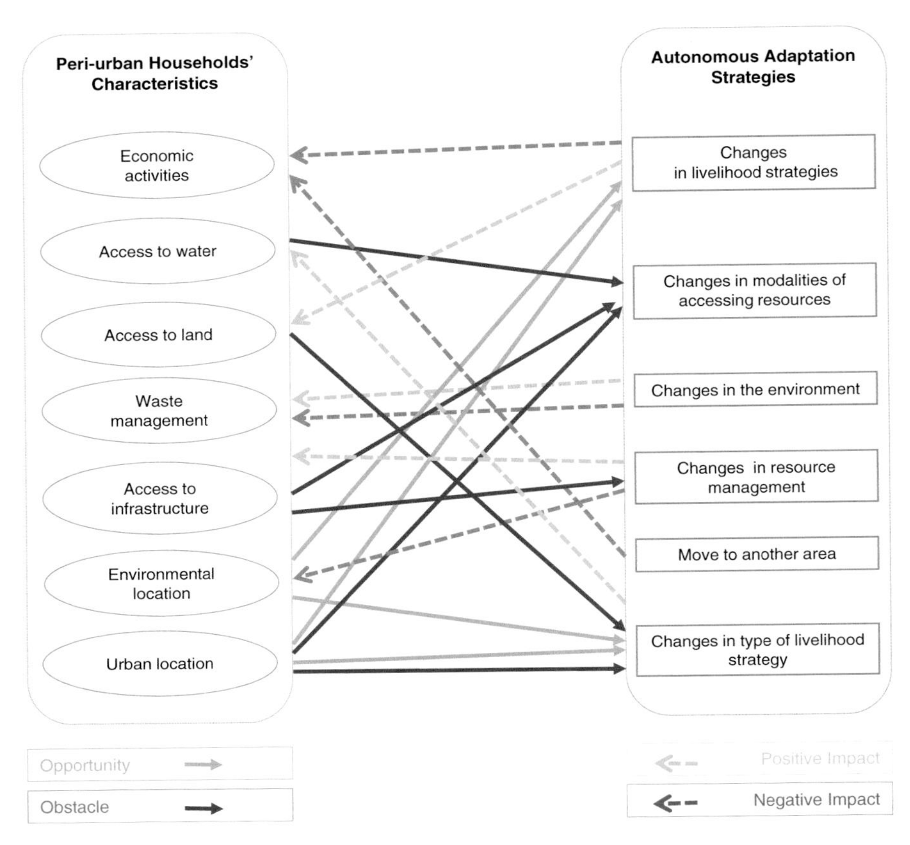

Fig. 5.13 Examples of possible opportunities or obstacles and negative or positive impacts deriving from the interaction between autonomous adaptation strategies and the characteristics of peri-urban households. *Source* The author

According to field observations, the greater opportunities/options that are present in peri-urban areas are primarily based on the fact that peri-urban areas straddle both urban and rural economies, and both constructed and natural environments.

First of all, the vicinity to urban areas allows for activities to be carried out that are not directly linked to natural resources, that can ensure livelihoods in the face of worsening environmental conditions, and that offer the possibility of adapting to environmental changes. Livelihood diversification is in fact recognized as an instrument for increasing what some have defined as the resilience of the socio-ecological peri-urban system (Foxon et al. 2008; Folke 2006) in confronting environmental changes, or, according to others, for reducing vulnerability.

Second, many of the identified adaptation options are linked to the coexistence of urban and rural features and activities that is typical of peri-urban areas, rather than the vicinity to the city. It is precisely this dependence on the characteristics of peri-urban households that allows for the evaluation of an adaptation practice. By relating an adaptation option to the qualities of peri-urban households, one can obtain four types of information: opportunities, obstacles, negative impacts and positive impacts (Fig. 5.13). The first two types evince how various modalities of resource management and access, location, services present, and economic activities can inhibit or promote a particular adaptation practice. Specifically, a given characteristic generates an 'opportunity' when it favours a particular autonomous adaptation practice (e.g. being located near large markets or major transportation routes facilitates commercial activities such as diversification of livelihood activities; being located near a river makes it easy to extract water from fairly shallow wells). 'Obstacles' remain, on the other hand, when a given characteristic inhibits or impedes an adaptation practice (e.g. being located near a transportation hub or major market may prevent the expansion of on-site agricultural land as a result of higher population density). The other two types of interaction (positive and negative impacts) demonstrate the kinds of effects that autonomous adaptation practices can have on the characteristics of peri-urban areas. In particular, a 'positive impact' occurs when an adaptation practice improves environmental or economic conditions (e.g. the use of fertilizer or the increase of land area for cultivation can improve waste management through recycling of organic waste, and thus has a positive impact). A 'negative impact' occurs when the adaptation practice introduces risks or stresses for natural resources (e.g. the construction of embankments across drying river beds in order to preserve water and use it for agriculture and domestic purposes could damage the river ecosystem and exacerbate the effects of changing climate on people (restricting access to water) and on natural resources).

Chapter 6
Conclusions: The Distance Between Critical Review and Institutional Commitment

Abstract The assumptions of prevailing interpretive approaches in the peri-urban sphere are discussed, and the research questions are reformulated in light of the analysis of Dar es Salaam. Seven key issues for a new interpretation of peri-urban areas and sub-Saharan cities are identified: relationship with natural resources, socio-economic and cultural homogeneity, environmental management and adaptive capacity, 'people as infrastructure', the 'ideal of life', dynamism in use of and access to resources, and rural–urban interdependence and bidirectional migration. Two sets of limits in research on peri-urban planning are thus identified. The first consists of the risk of formalizing, paralyzing or constraining social relations and 'platforms of action' within a structure that is incapable of responding to the needs of peri-urban residents. The second involves the risk that the emphasis on agency and informality in African cities become a vicious circle of self-exploitation and poverty that precludes development alternatives. Finally, the renewal of interpretive approaches in the cities of the Global North is envisioned together with a reflection on the need to bridge the gap between knowledge and planning.

Keywords Hybridization · Social network · Equity · Self-exploitation · Rigidity · Diversification · Platforms of action

One result of this research process is the identification of the main characteristics of peri-urban areas and their relationship with adaptive capacity to environmental changes. As emerges from some of the representations proposed in the literature, peri-urban areas have hybrid characteristics, where urban and rural features coexist and meld. Hybridization encompasses many aspects, from socio-economic characteristics to the modalities of natural resource use, from the morphology of settlement to formal and informal mechanisms of land governance. In this framework, the qualities defined as rural do not represent a residual and wasted aspect, a memory of a rural mode that is completing its transformation into urban, rather they are a structural component that is tightly interweaved into the urban dimension. It is through the rural–urban relationship and interdependence that another modality is established, one which is neither urban nor rural, and which produces local

L. Ricci, *Reinterpreting Sub-Saharan Cities through the Concept of Adaptive Capacity*, SpringerBriefs in Environment, Security, Development and Peace 26, DOI 10.1007/978-3-319-27126-2_6

economies, lifestyles, physical characteristics, and environmental management practices.

Based on this definition, which is one of the more commonly accepted of the variety of definitions found in the literature, the analysis of the peri-urban areas of Dar es Salaam and the search for key factors in environmental change adaptation has allowed for a reconsideration and critical discussion of several assumptions at the basis of planning policies and processes, in sub-Saharan cities and elsewhere. Such assumptions are the product of interpretive approaches informed by what, in the first chapter, has been defined as 'asymmetrical ignorance', approaches which apply models, categories and solutions developed in and for the Western city, in contexts that are very different from an environmental, social, cultural, and economic perspective.

The persistence of these interpretive approaches is determined by the cultural and scientific monopoly of the West, which leaves no room for forms of spatial governance based on principles that are not closely related to economic growth as an antidote to environmental degradation, inequality, and poverty. The pressures placed by the environmental transformations linked to global environmental change amplify and highlight the inadequacy of interpretive approaches, related policies and the actions adopted. The risks inherent in the recourse to interpretive approaches that are so detached from local reality, and the consequent imposition of inadequate development models, is that they exacerbate the impacts of climate change.

How to construct an interpretive approach for peri-urban sub-Saharan Africa that is free of both excessive Afro-pessimism, which is concentrated on the lack of 'modern' urban characteristics, and Afro-optimism, which exalts the creativity of the practices and capacities for self-organization that are location-specific, is an open question, one to which the reconsideration of certain assumptions that have guided public intervention in peri-urban areas to date might contribute. In the cities of sub-Saharan Africa and elsewhere, this means concentrating on modalities of informal urbanization, as well as the processes of hybridization between urban and rural activities and forms, processes that characterize the majority of African cities and constitute the reality of urban life.

One question that guided the exploration of peri-urban areas in Dar es Salaam was therefore whether to promote and support a transition to the urban mode that was deemed inevitable and desirable, and thus the only and best way to manage these areas, neither urban nor rural, and to reduce their vulnerability to environmental changes and other stresses.

Field analysis, as well as some of the literature, particularly that of post-colonial origins, seems to suggest that this probably is not always the case. Transition to the urban mode is stimulated by external factors that do not correspond to the expectations, needs, and life choices of the people who live in the city. Moreover, accompanying this transition through the development of Western types of infrastructure and services often means ignoring, breaking or modifying social networks, social capital, and the connected capacity to manage the environment and resources. The question, therefore, is to what point the system of 'platforms of action', through which people who live in peri-urban areas confront the difficulties and challenges posed by climate change and other factors, can be substituted, and what their substitution would entail in terms of environmental, social and economic sustainability.

The contemporary framework of policies, theoretical-disciplinary approaches and the reality of urban life, pose the unavoidable challenge of how to conceive of a knowledge 'project' that balances a critical interpretation of power dynamics, physical-environmental and social dynamics, and the commitment to ideating and generating effective institutional action, a project that begins with consideration and observation of what actually exists and occurs in peri-urban areas, rather than what is missing, in order to arrive at a new formulation of the problem.

6.1 Reinterpreting Areas with Hybrid Rural–Urban Characteristics in Cities of Sub-Saharan Africa

The first major contributions of this research are the points, discussed briefly here, which have led to a reconsideration of several assumptions contained in the prevailing interpretive approaches used in peri-urban areas and sub-Saharan cities. The considerations that emerge from the analysis constitute the elements of an interpretive approach for the cities of sub-Saharan Africa that will be useful for planning actions for climate change adaptation, an approach based on the ideas offered by the South African school of post-colonial studies (Simone, Robinson, Pieterse, Simon, Murray and Myers), and highlights the manner in which those elements allow for a reinterpretation of what occurs in peri-urban areas in relation to the evaluation of adaptive capacity.

6.1.1 Relationship with Natural Resources

The relationship with natural resources is a crucial issue, a key element for the people who live in peri-urban areas. While many studies have addressed households' economic dependence on access to such resources, and their role as a means of subsistence, their cultural and social values have been the subject of little inquiry. How natural resources contribute to sustaining people's livelihoods, and through what forms of social organization and society–nature relationship, is still relatively unexplored. It is not only economic necessity, but also the choice of a particular lifestyle (rural–urban) that defines a relationship with natural resources that is tightly linked to social networks and the capacity to manage those same resources.

Access to resources, like many activities in peri-urban areas, occurs through predominantly informal modalities, and as a result it is considered fairly insecure and a source of vulnerability. Nevertheless, the distinction between formal and informal is blurred: activities like peri-urban agriculture can play a role in the subsistence of households with an integrated income-generating aspect, thus highlighting both the importance of diversifying economic activities and the delicate equilibria of environment and social relations upon which the vulnerability of individuals depends. Water supply, for example, on which many economic activities, including agriculture, depend, is strongly influenced by environmental transformations, by social shifts that might be caused by political or economic choices, and by changes in culture or in spatial configuration of peri-urban areas; these factors can actually eliminate or modify 'platforms of action' and the relations that water supply depends upon, impeding or limiting some people's opportunities to access water.

When planning options for the future, an interpretive approach that looks at the interdependence between economic and socio-relational aspects leads one to also consider the cultural and social importance of relationships with resources that go beyond 'reasons of survival'. In this perspective, it is realism (rather than pessimism) to acknowledge that providing infrastructure for more than two-thirds of the city (the portion that is peri-urban) is not only a utopian plan in terms of financial resources and environmental costs, but would also mean constructing a city that is not there for people who do not exist. For such people, the 'urban' existence and the concurrent dependence on natural resources is not a residue of rurality, resulting from contingent needs, but a modality of the production of space that cannot be substituted with another in which the relationship with natural resources is managed through a 'technological filter'.

6.1.2 Socio-economic and Cultural Heterogeneity

Another question is immediately associated with that of access to resources: the socio-economic and cultural heterogeneity that characterizes the peri-urban areas of sub-Saharan Africa.

The socio-economic and cultural heterogeneity of people who live in peri-urban areas became quite clear during the questionnaire analysis and discussions with local civil servants, and it is also documented in the literature (Tacoli 1998, 2003; Mattingly 1999; Allen 2003). Nevertheless, it should be noted that most of the studies produced so far have emphasized economic elements related to systems of access to land and granting of land little, while heterogeneity was rarely analyzed with respect to the relationship with natural resources. In other words, it is often taken for granted that dependence on natural resources is a form of society–nature relationship that the poor develop out of necessity. This excludes the possibility that it is a lifestyle choice, one which implicates systems of relationships between people or groups and various, complex, and heterogeneous platforms of action, which are not created and used exclusively by 'subordinate' groups that support each other in order to survive. The main problem relative to this aspect is that, although part of the literature has documented the heterogeneity of the characteristics of peri-urban areas in the global South and North, too often peri-urban areas are generalized and assimilated with 'slums' or shantytowns that are slightly less dense but which have the same economic, social, and environmental problems. This vision creates several dangerous distortions at the level of public action, leading, on the one hand, to emergency interventions, sanitization, demolitions, and transfer of populations, while on the other, to blindness to the opportunities that peri-urban areas offer for sustainable urban development.

6.1.3 Environmental Management and Adaptive Capacity

On the other hand, the capacity for environmental management and adaptation to environmental change is closely correlated to the relationship that people have with natural resources, through what have been defined as 'platforms of action'. One of the questions that has guided this research process is directed at understanding whether peri-urban areas, and specifically those in sub-Saharan cities, are in fact more vulnerable than urban areas while lacking the means and the resources to face the contemporary critical environmental and economic situation. An elevated capacity for environmental management and adaptation has been identified, which results from infrastructure made by people and which allows them to live (not merely to survive) and maintain what one might call 'resilience'. The ability to observe the natural 'symptoms' and effects of environmental change, a profound understanding of natural cycles, together with the possibility of diversification, all of which persist in peri-urban areas, represent fundamental components of this capacity. Strategies, such as self-construction of dykes or changing agricultural practices according to changes in rain patterns, allow for a reduction of the impacts linked to environmental change. Nevertheless, one must not forget that these are partially determined by the limited capacity of institutions to intervene, and may constitute in some cases what is defined as 'maladaptation', which leads to a variety of negative impacts. Since they are not considered by institutions and are managed in an autonomous manner, they actually run the risk of transforming into practices that are damaging to the environment or that generate imbalanced and unjust power relations. The usefulness of a constant interaction between institutions and people in avoiding such problems must therefore be underlined.

The relationship with natural resources and the hybrid character of peri-urban areas allows people to diversify their economic activities and their modalities of access to resources, and this renders them much less vulnerable, given their various adaptation 'options' as compared with people living in urban areas.

Authors working in a variety of fields (ecology, political economy, agrarian science) maintain that 'diversity' and 'diversification' are means of coping with environmental shock and stress, constructing adaptive capacity (Holling 2001; Levin et al. 1998), and stimulating innovation and collective learning (Olsson et al. 2006; Ostrom 2005; Frenken 2004). A system's level of diversity is considered one of the principal characteristics of resilience (Gunderson/Holling 2002), and therefore of corresponding vulnerabilities and adaptive capacities. According to these approaches to sustainability management, one must not push beyond the limits of an urban system, but rather maintain and develop its variability and diversity, which means maintaining its adaptive capacity (Berkes et al. 2003).

If, on the one hand, the importance of diversification is recognized, on the other, there is a clear push, in strategies past and present, for completion of the urbanization process, formalization, and infrastructure. Having observed how the peri-urban areas of Dar es Salaam and most other sub-Saharan cities are conceived as unplanned areas to be rehabilitated or completed on various levels, it is important to stress not only the value of diversification, but also its dependence on the rural–urban interaction and how that dependence calls into question the imperative of urban transition. The possibility of having a variety of options for subsistence and adaptation is based on the presence of hybrid rural–urban forms and activities and temporary land use that would be eliminated by completion of the urban transition. This does not mean that the transition should be avoided in order to maintain a state of equilibrium (precarious as it is), but rather that a process that accelerates that transition as an instrument for improving adaptive capacity (as well as environmental security and economic development) must take into account what is present and what will be eliminated, not merely what will be added.

6.1.4 'People as Infrastructure'

When considering what is present in peri-urban areas and the complex system of organization of people for environmental management and distribution of access to resources, the inherent risk of a negative vision concentrated exclusively on what peri-urban areas lack in order to be urban becomes clear, as does the importance of 'unconventional' infrastructure and services. By using the concept of 'people as infrastructure' formulated by Simone (2004), one can characterize perhaps the most important part of peri-urban areas, which too often remain in the shadows. This provides a cognitive framework with which to respond to the vulnerability of peri-urban areas and the modalities with which difficulties can be addressed, since it recognizes people, as individuals or groups, as protagonists of action and spatial transformation, with the knowledge and capacity to act in a meaningful and strategic manner. This agency, which is also expressed through management of resources and the environment, is a determinative factor of adaptive capacity. It is fundamental that agency be considered an integral part of the spatial production process, thereby overcoming the negative vision of lack of knowledge, competence, means, and capacity. This leads to recognition of a system of power and relational knowledge, as well as the capacity to interact with the environment, to work in one's own interest. As actors who are conscious of spatial transformation, people actively and strategically interact with their environment through a continuous process that is constantly revised and recreated when circumstances change, interests merge or diverge, and objectives change. Entering into a relationship of 'strategic' connection with other people or networks of people (which might mean guaranteed access to water or land) allows a person to negotiate and defend his or her position, and to safeguard resources and relationships that have specific value.

How planning can be inspired by these 'platforms of action' remains to be investigated. In particular, we need to understand whether and to what degree they can inform a project of infrastructural systems that substitute or integrate modern networks of steel and cement, or hybridize with them to better adapt to the peri-urban context, rather than destroying the system of relations upon which people depend.

6.1.5 The 'Ideal of (Urban) Life'

Another element that leads to an examination of the interpretive approaches for peri-urban areas arose following the comparison of policy views that emerged from institutional documents and interviews, and from questionnaires and observations of peri-urban areas. In the former, transformation of peri-urban areas is yet again supported in order to elevate them to urban status, in the name of equal access to shelter and adequate space, comparable to a modern and western city 'such as New York'. In the latter, the answers to questionnaires reflect opinions and behaviours of people who demonstrate the opposite tendency. None of the people living in peri-urban areas claimed to want to move to an area with more urban characteristics; on the contrary, many of the people interviewed in the urban area said that they wanted to move to peri-urban areas. The presence of hybrid spaces in which to undertake both urban and rural activities therefore represents a characteristic that is important for the people of Dar es Salaam.

The fact that this is attributable not only to subsistence needs is evinced by the observation of a widely discussed phenomenon, upon which the criticism of upgrading programmes in the South is often based. Specifically, in Dar es Salaam, as in other cities, the outcome of upgrading programmes for informal consolidated settlements (for example, the *Community Infrastructure Upgrading Programme* implemented in Manzese, in the Kinondoni municipality) is the transfer of people to peri-urban areas. This occurs for a variety of reasons: those with formal land title seize the opportunity to sell their plot, whose value was increased through the upgrading programme (this allows them to buy another larger plot in a peri-urban area); people who do not have formal land title to the plot they occupy, or possibly rent, are forced to move without any benefit from the sale of their plot; the pressure for urbanization and the costs entailed by the regularization and formalization of land tenure cannot be reconciled with individuals' or households' subsistence strategies. This raises another important question related to the importance that is placed on security of access to and use of resources (such as water, land, etc.) and their relationship with dynamism and diversity in the informal forms of granting rights.

6.1.6 Dynamism in the Use of and Access to Resources

Access to resources and services represents a fundamental need that has always guided policies and programmes for urban development in the Global South, and also represents the basis upon which adaptive capacity is constructed within the debate on climate change. What the elements are that guarantee that access remain widely controversial and debated. What instruments planners and politicians can use to guarantee that people have equal access to land, housing, water, energy, and other resources is a complex question that requires elaboration, especially as regards peri-urban areas, since the continued use of non-institutional modalities of regulation and legitimization conflicts with policies and interventions imposed from above.

In particular, 'access to land'[1] has long dominated development, poverty reduction, and food security policies, and, in recent years, reduction of vulnerability to climate change. It is widely believed that there is a close relationship between access to land and formalization of use and occupation rights, between regularization and security of access to land as a productive asset or living space. Nevertheless, the phenomenon of resident transfer that occurs following upgrading programmes and granting of formal land tenure, demonstrates that this relationship is not guaranteed. It has been noted that in Dar es Salaam a majority of the people do not have a title deed to occupy the area in which they reside, and that the modality through which buying and selling occurs follows processes that are almost always formal–informal hybrids. Despite the fact that there is a strong push for formalization (see Sect. 3.3.2), many people, independent of their economic position, choose to continue using an informal, more flexible regime. If, as has been asserted, there is a need or a will to live in areas where it is possible to diversify sources of income, where there is access to natural resources, and even the opportunity to diversify forms of water supply (for example), formalization processes paralyze this system and in many case render it less 'secure', more exclusive and certainly less flexible, with all the positive and negative consequences that result therefrom.

Moreover, the concrete outcomes of the liberalization process initiated with the Structural Adjustments, and of the current push for formalization demonstrate that many people are still very wary of becoming involved in what is presented as an opportunity, such as the possibility of possessing legal title deed (Briggs 2011; Mwanfupe 2007). The land formalization process has different results for different segments of the population, and the advantages offered by formalization seem to be greater for the people who are wealthier. Most of those people are already included in the employment taxation system for the public sector and the formal private sector; having already been identified by the Treasury, they are less cautious and concerned than the people who are still invisible to the State. The wealthy also have

[1]Here 'access to land' is intended as the possibility of using a space, whether privately or publicly, individually or collectively, for settlement and in order to carry out economic or other activities.

greater access to capital, from both savings and credit, which allows them to actively participate in the real estate market, while the poorer tend to be increasingly marginalized when the real estate market rises and they decide to sell their land as a result. Paradoxically, it would therefore seem that the process for granting title deeds for land use and occupation in Dar es Salaam tends to be of greater benefit, in the medium and long term, to urban residents who are already rich, thus reinforcing economic inequality within the city.

In other words, we find ourselves faced with a widespread conviction that the acquisition of formal title deed for land use or ownership will contribute to an increase in the security of the owner who invests in his or her land. But is the granting of formal titles still the best approach to guaranteeing access to land? Based on the analysis, the land title granting process might not be the only viable option. One must consider that the possession of a title deed could ensure credit only if the credit is available and if the title is accepted as a guarantee, and this is not necessarily the case. Moreover, the same process with which titles are granted could seriously alter other rights that people seek to protect, for example where the registration (and regularization) process would impede the fluid system of relationships in the traditional land tenure regime, or where the rights of third parties (people who occupy or use the land through recognition of second or third level indirect rights) are excluded in the name of total security of the principal title holder. Other ways of obtaining security of land use and/or occupation could also include people who have small and medium-sized plots, as is often the case in peri-urban areas. It seems to be political legitimization at the local level rather than certification of a formal title deed that contributes to guaranteeing a right of access to land; if people have faith in the mechanisms of land distribution (or mechanisms that render land available or allow for negotiation of acquisition processes) or conflict resolution mechanisms related to land, they will feel more protected (more secure) and therefore more disposed to invest in their own land. This does not mean that a clear and juridical definition of rights is not important. The two questions are, in fact, closely related. But too often the emphasis is placed exclusively on the types of rights and documentation that are granted, and not enough on the processes for assuring such rights and the mechanisms for defending and maintaining them. What probably requires further research, before the regularization process, is therefore the relationship between the local informal level of political legitimization and the system of regularization for formal possession. The question is, therefore, if and how planning can conceive of intermediate regulatory systems, in which informal processes can also be expressions of 'sovereignty' and various forms of legitimization can be recognized, both for use of and access to land. Such forms of legitimization could also be used for other resources, and that which is not legal could still be recognized as legitimate, something that in fact already occurs. Recognizing that legitimacy also allows for management of the negative aspects that derive from the impossibility of monitoring certain informal mechanisms and power relations (or from the desire to avoid monitoring them in a transparent manner) that could lead to situations of oppression or exploitation and environmental risk.

6.1.7 Rural–Urban Interdependence and Bidirectional Migration

Another aspect that has sparked a variety of debates is the relationship of interdependence between urban and rural areas. While the old theory of urban parasitism and the exploitation of the hinterland by the city seems to have been overcome in certain senses, there remains a shared belief that peri-urban areas are the destination of migrants from rural areas who settle there temporarily because the cost is lower and because it offers the possibility of producing their own food until their wages earned in the urban area are sufficient for a transfer into a better neighbourhood with infrastructure and services. Many elements discovered in Dar es Salaam suggest that this cannot be adopted as the principal model of rural–urban relations.

In the first place, though migration from rural areas is certainly motivated by the search for the better opportunities offered by urban areas, this does not coincide with what Mattingly (2009) would define as the inevitable transition to urban. In other words, the transfer from rural areas is often aimed at non-temporary settlement in a peri-urban area because it is a place where one can benefit from both rural and urban opportunities while maintaining a good level of flexibility, an important resource, particularly when faced with environmental pressures.

Moreover, many transfers occur from the centre of the city towards peri-urban areas, in response to the pressures of urbanization and formalization processes and for the advantages of living in semi-natural environments, such as the opportunity to diversify means of subsistence, a higher quality of life and the connection with rural culture and traditions.

On the other hand, among the people who move from rural areas towards the city, the poorest are likely to settle in deteriorating and congested areas (slums) near the centre of the city, where they are not forced to bear the cost of transportation and are able to find temporary work more easily.

In the planning process, an understanding of the interdependencies and relationships between urban and peri-urban areas is crucial, because it informs the definition of use of space practices, as well as the rhythm of movements (daily, weekly or less frequent) and transfers of residence between different areas. These processes call attention to the dynamism of peri-urban areas, to the rapid and continuous process of transformation attributable, in part, to the pressure of urbanizations processes, and to the practices people use to sustain their own livelihoods and confront environmental change.

6.2 Responding to Asymmetrical Ignorance: Autonomous Adaptation as an Opportunity and a Trap

These considerations lead to several conclusive reflections on the relationship between modalities of production of space in peri-urban areas, adaptive capacity to environmental change and institutional planning strategies for urban development.

The case of Lagos illustrates how African cities are characterized as spaces and practices that represent a challenge for contemporary urban thinking, or rather, an opportunity to rethink and rework our knowledge of cities, not only in Africa but in the Western world as well. That is not to suggest that cities like Lagos and Dar es Salaam are a model for future urban development, given their capacity to function despite their apparent lack of means, infrastructure, coordination and planning. To do so would be to risk ignoring several environmental and social criticalities of such cities and condemn a part of the population to a state of distress and difficulty. Rather, their contribution to an interpretative approach that takes into account the questions outlined above lies in the recognition that these cities are defining their own vision of African urbanization. That experience, even if only temporarily, could open an authentic dialogue between cities of the South that are rapidly growing, and therefore bring the African city to the centre of decision-making processes and political debates, contributing at the same time to an enrichment of the Western perspective on urban development.

Lastly, by bringing together the reflections on peri-urban areas and adaptive capacity, two main limits are highlighted that are at the heart of the questions discussed above, limits which derive from that lack of consideration for the specific characters of peri-urban areas—and above all from the lack of consideration for what we have defined as 'platforms of action'—or, on the contrary, from glorifying the creativity and self-organization of African cities.

First, we can define a '*rigidity limit*'. It has been noted that normative and prescriptive processes, such as formalization of land tenure, can represent an obstacle for livelihood diversification and access to resources, but then can also bring an end to phenomena of exclusion. For that reason, interpretive approaches that explore and take into account the modalities of informal legitimization and management of resources, can contribute, on the one hand, to the emergence of alternative modalities of management and access to resources that enhance the 'platforms of action' identified in peri-urban areas (thus maintaining the advantages that derive from the presence of an informal system and flexible access to and management of resources). On the other hand, they can contribute to controlling the negative effects of those platforms on the environment and people, and to understanding their role in people's subsistence strategies.

The second limit is that of '*self-exploitation*', which derives from the glorification of the capacity of African city dwellers to act in an autonomous informal and effective manner.

From the ever-expanding research on African cities come perhaps the most complex articulations of agency and subaltern subjectivity. On one hand, this literature focuses on people's heightened capacity to utilize 'registers of improvisation' in moments of crisis (Mbembe/Roitman 2003: 114, in Roy 2009: 826–827), for which informality becomes a modality of expressing their own subjectivity, an effective method of acting in cities with few resources. In this perspective, modern steel and cement infrastructure is substituted with social networks' platforms of action (Simone 2004). On the other hand, one asks whether the modalities of autonomous organization, environmental management and adaptation are actually an effective collective system of production of space, or rather a 'desperate search for human agency (improvisation, incessant convertibility) in the face of a neoliberal grand slum' (Watts 2005: 184). That which was first open and flexible becomes, from this perspective, self-exploitation, the trap of poverty and oppression (Watts 2005).

Thus, we find ourselves faced with a dualism between the need to 'subvert' an urban development model guided by global urbanization processes (neo-liberalism) that do nothing but reproduce self-exploitation of subalterns (Watts 2005), and the glorification of local and informal practices as a way of 'operating more resourcefully in under-resourced cities' (Simone 2006: 357).

How to break away from this dualism is certainly a complex question that several authors have examined (Roy 2009). It requires, however, reference to specific cases and contexts of observation. These limits call into question the planning policies and practices not only of African cities, but of Western cities as well. The phenomena of hybridization and self-organization are not, in fact, absent in cities of the North, an observation which reopens the historic debate on urban sprawl as a model of negative development and spontaneous settlement practices.

This last point leads one to reflect on how planning can be based on the operations of power and control exercised through formalization measures as a normative concept of the city, which have an impact on people's lives and their right to access resources; or alternatively, how it can refer to other systems of recognition of rights that might better include questions of vulnerability. In other words, one wonders how to confront ordinary situations of conflict or vulnerability, such as those linked to environmental change, without relying on rigidly structured processes, technical–bureaucratic instruments and 'legalization', but rather by redefining the challenges related to rights and attributions, to the capacity to act and to be a part of decision-making processes (Roy 2005: 150).

Phenomena that were once considered unique to cities of the South are now being recognized in various cities in the North as well. The question of informality and its relationship with the environmental changes underway, urban rehabilitation and upgrading programmes, the introduction of insurance mechanisms for reducing the economic impacts of extreme climatic events that are based on private property, all clash with the old community-based customary land tenure systems and mutual aid. Conflicts over the settlement modalities of the Roma and other groups of people who use alternative modalities for land access and resource use, equally present though certainly more hidden in Western cities, reignite tensions between formal and informal land use and resources even in Western cities, and could initiate a reconfiguration of property rights, access to resources, and services.

With respect to the question of urban sprawl and hybrid rural–urban forms viewed as a model of unsustainable development, these reflections on African cities could lead to a reconsideration, from a positive perspective, of the ecological functionality of such areas (linked to the discontinuity in soil impermeabilization, which has positive effects on natural resource cycles and prevents climatic distortions of the city, and to low population density, which reduces the concentration of greenhouse gases emissions), but also the network of social and spatial relations that are constructed in such spaces, and the value that they have even in the Western context, which appears to be little dependent on rural–urban linkages and interdependencies. Moreover, urban infrastructure and the dense city are displaying their fragility in the face of increasingly frequent and destructive extreme climate events.

It is that fragility that prompts a reconsideration of the relationship between urbanization and economic growth, which has been widely studied and often used to justify the expulsion of hybrid or rural activities from the city and the preference for compact and polycentric cities, models of urban development associated with a lower level of vulnerability and environmental change. This relationship may not be a given, as several authors have pointed out (Polese 2005). It is difficult to demonstrate that cities (urban agglomeration in itself) generate economic growth and a capacity to respond positively to environmental change. Both depend, in fact, on many factors, including the complex dynamics among which we can insert people's capacity to act and social networks, not merely goods, infrastructure, technology or financial resources. Cities are therefore important, not because they are the only motor of economic growth, but because people live increasingly in urban areas, and it is there that their economic activities take place. If the triangular relationship (of causality) between economic growth, reduction of vulnerability to environmental change, and urban development is not assumed a priori, even transition processes of peri-urban areas towards an urban 'infrastructured' and 'formalized' state require reconsideration, and the other possible paths that the people who live in these areas are perhaps already using should be explored.

6.2.1 Bridging the Gap Between Knowledge and Planning

The knowledge provided through the analysis of peri-urban areas in Dar es Salaam and the critical reflection on unrealistic planning assumptions should support a radical overhaul of inadequate formal and modernist planning to effectively address the new challenges posed by environmental and climate change. The benefits and burdens of autonomous adaptation are crucial in decision-making for political and social sustainability and in identifying local actions for the adoption of more inclusive and participatory mechanisms.

The growing concerns regarding equity, fairness and justice in adaptation and urban planning adaptation choices highlight that adaptation, as well as improvement of local adaptive capacity, calls for policy change.

Autonomous adaptation practices generate direct and indirect impacts at different scales (community, local, regional, national, and international) and reflect unequal access to resources underpinned by political relationships within communities and between them and the broader political and economic system. Understanding those relationships is crucial in policies and planning interventions for adaptation which may reinforce or exacerbate existing inequalities and dynamics associated with climate change effects.

Therefore, the above-mentioned elements for reinterpreting sub-Saharan cities can support the integration of societal differentiation in livelihoods, resource access, and resultant climate risk, in policy and planning processes avoiding possible conflicts between adaptation, sustainable development, and poverty reduction. In this perspective, adaptation and urban and environmental policies and planning are not only a response to environmental change and urban growth challenges, but a set of individual and collective political choices within an institutional framework. Therefore, effectively integrating adaptation and development would require going beyond the mere 'climate-proofing' of development sectors, and addressing institutions, governance, and political discourses through transitional and transformative forms of adaptation.

References

Adell, G., 1999: "Theories and Models of the Peri-urban Interface: A Changing Conceptual Landscape", Paper Produced for the Research Project on Strategic Environmental Planning and Management of the Peri-urban Interface, Development Planning Unit, University College London.

Adger, W.N., 1999: "Social Vulnerability to Climate Change and Extremes in Coastal Vietnam", in: *World Development,* 27,2: 249–269.

Adger, W.N.; Paavola, J.; Huq, S.; Mace, M.J., 2006: *Fairness in Adaptation to Climate Change* (Cambridge, MA: MIT Press).

Adger, W.; Vincent, K., 2005: "Uncertainty in Adaptive Capacity. (IPCC Special Issue on 'Describing Uncertainties in Climate Change to Support Analysis of Risk and Options')", in: *Comptes Rendus Geoscience,* 337,4: 399–410.

Alam, M.; Rabbani, M.D.G., 2007: "Vulnerabilities and Responses to Climate Change for Dhaka", in: *Environment & Urbanization*, 19,1: 81–97.

Alber, G.; Kern, K., 2008: "Governing Climate Change in Cities: Modes of Urban Climate Governance in Multi-level Systems [online]", OECD Conference Paper, Competitive Cities and Climate Change, 9–10 October, Milan, Italy.

Alberti, M., 2009: *Advances in Urban Ecology: Integrating Humans and Ecological Processes in Urban Ecosystems* (New York: Springer).

Allen, A., 2010: "The Green vs Brown Agenda in City Regions, Peri-urban and Rural Hinterlands", Inception Workshop Oxford Brookes University 18–19 Maggio 2010.

Allen, A., 2001: "Urban Sustainability Under Threat. The Industrial Restructuring of the Fishing Industry in the City of Mar del Plata, Argentina", in: *Development in Practice (May),* 11, 2–3: 152–173.

Allen, A., 2006: "Understanding Environmental Change in the Contest of Rural-Urban Interaction", in: Gregor, M.; Simon, D.; Thompson, D. (Eds.): *The Peri-urban Interface. Approaches to Sustainable Natural and Human Resource Use* (London: Earthscan).

Allen, A.; You, N., 2002: *Sustainable Urbanization: Bringing the Green and Brown Agenda* (London: Development Planning Unit/UN-Habitat/DFID).

Allen, A.; da Silva, N.; Corubolo, E., 1999: "Environmental Problems and Opportunities of the Peri-urban Interface and Their Impact upon the Poor", Paper Produced for the Research Project on Strategic Environmental Planning and Management for the Peri-urban Interface, Development Planning Unit, University College London.

Al-Sayyad, N.; Roy, A. (Eds.), 2004: *Urban Informality: Transnational Perspectives from the Middle East, Latin America and South Asia* (Boulder, CO: Lexington Books).

Abrahamsen, R., 2000: *Disciplining Democracy: Development Discourse and Good Governance in Africa* (London: Zed Books).

Amin, A.; Graham, S., 1997: "The Ordinary City", in: *Transactions of the Institute of British Geographers,* 22,4: 411–429.

L. Ricci, *Reinterpreting Sub-Saharan Cities through the Concept of Adaptive Capacity*, SpringerBriefs in Environment, Security, Development and Peace 26, DOI 10.1007/978-3-319-27126-2

Annez, P.; Buckley, R.; Kalarickal, J., 2010: "African Urbanization as Flight? Some Policy Implications of Geography", in: *Urban Forum,* 21,2010: 221–234.

Appadurai, A., 1996: *Modernity at Large: Cultural Dimensions of Globalization* (Minneapolis: University of Minnesota Press).

Argawala, R., 1983: *Price Distortions and Growth in Developing Countries. World Bank Staff Working Paper No. 575* (Washington D.C.: The World Bank).

Armstrong, A., 1986: "Colonial and Neocolonial Urban Planning: Tree Generation of Master Plans for Dar es Salaam, Utafiti", in: *Journal of Faculty of Arts and Social Science* (University of Dar es Salaam), VIII,1: 43–66.

Armstrong, A.M., 1986: "Urban Planning in Developing Countries: An Assessment of Master Plans for Dar es Salaam", in: *Singapore Journal of Tropical Geollraphy,* 7,I: 12–27.

Atkinson, A., 1992: "The Urban Bioregion as a 'Sustainable Development' Paradigm", in: *Third World Planning Review,* 14,4: 327–354.

Auld, H.; MacIver, D., 2005: *Cities and Communities: The Changing Climate and Increasing Vulnerability of Infrastructure.* An Extract from the Book Climate Change: Building Adaptive Capacity, Meteorological Services of Canada, Environment Canada.

Balbo, M. (Ed.), 2005: *International Migrant and the City* (Nairobi: UN-Habitat, Università Iuav di Venezia).

Barkan, J.D., 1994: *Beyond Capitalism Vs. Socialism in Kenya and Tanzania* (Boulder: Lynne Rienner Publishers).

Bateson, G., 1976: *Verso un'ecologia della mente* (Milano: Adelphi) (ediz. orig. 1972).

Beall, J.; Fox, S., 2009: *Cities and Development* (London: Routledge).

Beall, J.; Guha-Khasnobis, B.; Kanbur, R., 2010: "Introduction: African Development in an Urban World: Beyond the Tipping Point", in: *Urban Forum,* 21: 197–204.

Beall, J.; Khasnobis, G.B.; Kanbur, R., 2010: "Introduction: African Development in an Urban World: Beyond the Tipping Point", in: *Urban Forum,* 21,3: 187–204.

Berrisford, S., 2014: "The Challenge of Urban Planning Law Reform in African Cities", in: Parnell, S.; Pieterse, E. (Eds.): *Africa's Urban Revolution* (London: Zed Books Ltd.): 167–183.

Bicknell, J.; Dodman, D.; Satterthwaite, D. (Eds.), 2009: *Adapting Cities to Climate Change: Understanding and Addressing the Development Challenges* (London: Earthscan).

Blaikie, P.T.; Cannon, T.; Davis, I.; Wisner, B., 1994: *At Risk: Natural Hazards, People's Vulnerability and Disasters* (London: Routledge).

Bologna, G., 2004: *State of the World 2004. Consumi. Invito alla sobrietà felice. Come vivere meglio consumando meno (a cura di G. Bologna)* (Milano: Edizioni Ambiente).

Bourdieu, P., 1977: *Outline of a Theory of Practice* (Cambridge and New York: Cambridge University Press).

Brennan, J.R.; Burton, A.; Lawi, Y. (Eds.), 2007: *Dar es Salaam: Histories from an Emerging African Metropolis* (Dar es Salaam: Mkuki ny Nyota Publisher).

Briggs, J., 2011: "The Land Formalisation Process and the Peri-urban Zone of Dar es Salaam, Tanzania", in: *Planning Theory and Practice,* 12,1: 115–153.

Briggs, J.; Mwamfupe, D., 2000: "Peri-Urban Development in an Era of Structural Adjustment in Africa: The City of Dar es Salaam, Tanzania", in: *Urban Studies,* 37,4: 797–809.

Briggs, J.; Mwamfupe, D., 1999: "The Changing Nature of the Peri-urban Zone in Africa: Evidence from Dar es Salaam, Tanzania", in: *Scottish Geographical Journal,* 115,4: 269–282.

Briggs, J.; Yeboah, I.E.A., 2001: "Structural Adjustment and the Contemporary Sub-Saharan African City", in: *Area,* 33, 8–26.

Bryceson, D.F., 1992: "Urban Bias Revisited: Staple Food Pricing in Tanzania", in: *European Journal of Development Research,* 4,2: 82–106. doi:10.1080/09578819208426572.

Bryceson, D.F., 2002: "The Scramble in Africa: Reorienting Rural Livelihoods", in: *World Development,* 30: 725–739.

Brown, R.L., 1980: *Il 29º giorno* (Firenze: Sansoni).

Budds, J.; Minaya, A., 1999: *Overview of Initiatives Regarding the Management of the Peri-urban Interface,* Draft for Discussion—Strategic Environmental Planning and Management for the Peri-urban Interface Research Project—Peri-urban Research Project Team, Development Planning Unit, University College London.

Bulkeley, H.; Betsil, M., 2005: *Cities and Climate Change: Urban Sustainability and Global Environmental Governance* (Routledge Studies in Physical Geography and Environment) (London and New York: Routledge).

Bulkeley, H.; Betsill, M., 2013: "Revisiting the Urban Politics of Climate Change", in: *Environmental Politics*, 22,1: 136–154.

Bulkeley, H., 2006: "A Changing Climate for Spatial Planning", in: *Planning Theory and Practice*, 7,2: 203–214.

Burton, I.; Malone, E.; Saleemul, H., 2004: *Adaptation Policy Frameworks for Climate Change: Developing Strategies, Policies and Measures* (Editors: Lim, B.; Spanger-Siegfried, E.), United Nations Development Programme (Cambridge, UK: Cambridge University Press).

Castells, M., 1983: *The City and the Grassroots* (Berkeley: University of California Press).

Chakrabarty, D., 2000: *Provincializing Europe: Postcolonial Thought and Historical Difference* (Princeton: Princeton University Press).

Chambers, R., 1989: "Editorial Introduction: Vulnerability, Coping and Policy", in: *IDS Bulletin*, 20,2: 1–7.

Cohen, B., 2004: "Urban Transition in Developing Countries: A Review of Current Trends and a Caution Regarding Existing Forecasts", in: *World Development*, 32,1: 23–51.

Cochrane, A.; Ward, K., 2012: "Researching the Geographies of Policy Mobility: Confronting the Methodological Challenges", in: *Environment and Planning A,* 44,1: 5–12.

Cutter, S.L.; Boruff, B.J.; Shirley, W.L., 2003: "Social Vulnerability to Environmental Hazards', in: *Social Science Quarterly,* 84,2: 242–261.

Dalal-Clayton, B.; Dent, D.; Dubois, O., 2002: *Rural Planning in Developing Countries: Supporting Natural Resources Management and Sustainable Livelihoods* (London: Earthscan).

Dar es Salaam City Council, 1999: *Strategic Urban Development Planning Framework* (Draft) (Dar es Salaam: Stakeholders Edition).

Dar es Salaam City Council, 2004: *City Profile for Dar Es Salaam, United Republic of Tanzania.*

Davila, J.D., 2002: "Rural-Urban Linkages: Problems and Opportunities", in: *Espaço e Geografia,* 5,2: 35–64.

Davis, M., 2006: *Planet of Slums* (London: Verso).

Coquery-Vidrovitch, C., 2005: "Introduction: African Urban Spaces: History and Culture", in: Falola, T.; Salm, S. (Eds.): *African Urban Spaces in Historical Perspective* (Rochester, NY: University of Rochester Press): xv–xl.

Davoudi, S.; Strange, I., 2009: *Conceptions of Space and Place in Strategic Spatial Planning* (London: Routledge).

Davoudi, S.; Crowford, J.; Mehmood, A., 2009: *Planning for Climate Change: Adaptation Mitigation and Vulnerability* (London: Earthscan).

De Soto, H., 2000: *Mystery of Capital: Why Capitalism Triumphs in the West and Fails Everywhere Else* (New York: Basic Books).

Dodman, D., 2009: "Blaming Cities for Climate Change? An Analysis of Urban Greenhouse Gas Emission Inventories", in: *Environment and Urbanization,* 21,1: 185–201.

Dodman, D.; Kibona, E.; Kiluma, L., 2011: "Tomorrow Is Too Late: Responding to Social and Climate Vulnerability in Dar es Salaam, Tanzania Case Study: Dar es Salaam", in: UN-Habitat, *UN-Habitat Global Report on Human Settlements 2011: Cities and Climate Change.*

Douglas, I.; Alam, K., 2006: *Climate Change, Urban Flooding and the Rights of the Urban Poor in Africa: Key Findings from Six African Cities* (London: Action Aid International).

Douglass, M., 1998: "A Regional Network Strategy for Reciprocal Rural–Urban Linkages: An Agenda for Policy Research with Reference to Indonesia", in: *Third World Planning Review,* 20,1: 1–33.

Dow, K., et al., 2006: "Exploring the Social Justice Implications of Adaptation and Vulnerability", in: Adger, W.N., et al. (Eds.): *Fairness in Adaptation to Climate Change* (Cambridge, MA: The MIT Press): 79–96.

Durand-Lasserve, A., 1998: "Rural–Urban Linkages: Managing Diversity. Governance as a Matrix for Land Management in the Metropolitan Fringes", International Workshop on Rural–Urban Linkages, Curitiba, Brazil, March 1998.

Ekers, M.; Hamel, P.; Keil, R., 2012: "Governing Suburbia: Modalities and Mechanisms of Suburban Governance", in: *Regional Studies,* 46,3: 405–422.

Eakin, H.; Luers, A.L., 2006: "Assessing the Vulnerability of Social-Environmental Systems", in: *Annual Review of Environment and Resources*, 31: 365–394.

Engle, N., 2011: "Adaptive Capacity and its Assessment", in: *Global Environmental Change*, 1–10.

Enwezor, O.; Basualdo, C.; Bauer, U.M.; Ghez, S., 2002: *Under Siege: Four African Cities.* Documenta 11: Platform 4. Ostfildern-Ruit: Hatje Cantz.

European Commission, 2006: *EU Action Against Climate Change: Helping Developing Countries to Cope with Climate Change* (Luxemburg: Office for Official Publications of the European Communities).

Fahmi, W., 2003: "City Inside Out(side): Postmodern (Re)presentations—City Narratives and Urban Imageries", Planning Research Conference 2003—Oxford: Oxford Brookes University web site, 8–10 April.

Faldi, G., 2010: "Valutazione della vulnerabilità al cambiamento climatico delle comunità costiere di Dar es Salaam (Tanzania) rispetto al fenomeno dell'intrusione salina nella falda acquifera" (M.Sc. Thesis).

Faldi, G.; Rossi, M., 2014: "Climate Change Effects on Seawater Intrusion in Coastal Dar es Salaam: Developing Exposure Scenarios for Vulnerability Assessment", in: Macchi, S.; Tiepolo, M. (Eds.): *Climate Change Vulnerability in Southern African Cities* (New York: Springer International Publishing): 57–72.

Falola, T.; Salm, S.J., 2005: *African Urban Spaces in Historical Perspective* (Rochester, NY: University of Rochester Press).

Florida, R., 2002: *The Rise of the Creative Class* (New York: Basic Books).

Folke, C., 2006: "Resilience: The Emergence of a Perspective for Social-Ecological Systems Analyses", in: *Global Environmental Change,* 16: 253–267.

Folke, C.; Carpenter, S.R.; Walker, B.; Scheffer, M.; Chapin, T.; Rockström, J., 2010: "Resilience Thinking: Integrating Resilience, Adaptability and Transformability", in: *Ecological Society,* 15,4: 1–9.

Folke, C.; Carpenter, S.; Elmqvist, T.; Gunderson, L.; Holling, C.S.; Walker, B., 2002: "Resilience and Sustainable Development: Building Adaptive Capacity in a World of Transformations", in: *Ambio,* 31: 437–440.

Fontanari, E., forthcoming: "Adaptation to Climate Change in the New Master Plan of Dar es Salaam", in: Macchi, S.; Ricci, L. (Eds.): *Adaptation Planning in a Mutable Environment: From Observed Changes to Desired Futures,* Research for Development Series (Switzerland: Springer).

Foxon, T.J.; Stringer, L.C.; Reed, M.S., 2008: *Governing Long-Term Social-Ecological Change: What Can the Resilience and Transitions Approaches Learn from Each Other? Long-Term Policies Governing Social-Ecological Change* (Berlin: Springer).

Freund, B., 2007: *The African City: A History* (New York: Cambridge University Press).

Friedmann, J.; Alonso, W., 1975: *Regional Policy: Readings in Theory and Applications* (Cambridge, Mass: MIT Press).

Friedmann, J., 1979: "Basic Needs, Agropolitan Development, and Planning from Below", in: *World Development,* 7: 607–613.

Friedmann, J., 1985: "Political and Technical Moments in Development: Agropolitan Development Revisited", in: *Environment and Planning D: Society and Space,* 3,2: 155–167.

Friedmann, J., 2005: "Globalization and the Emerging Culture of Planning", in: *Progress in Planning,* 6: 183–234.

Füssel, H.-M., 2007: "Adaptation Planning for Climate Change: Concepts Assessment Approaches, and Key Lessons", in: *Sustainability Science,* 2: 264–275.

Gallopin, G., 2006: "Linkages Between Vulnerability, Resilience and Adaptive Capacity", in: *Global Environment Change,* 16: 293–303.

Gandy, M., 2006: "Planning, Anti-Planning and the Infrastructure Crisis Facing Metropolitan Lagos", in: *Urban Studies,* 43,2: 371–396.

Gandy, M., 2005: "Learning from Lagos", in: *New Left Review,* 33: 36–52.

Graham, S.; Marvin, S., 2001: *Splintering Urbanism: Networked Infrastructures, Technological Mobilities and the Urban Condition* (London and New York: Routledge).

Grønlund, B., 2007: "Some Notions on Urbanity", Proceedings, 6th International Space Syntax Symposium, Istanbul, 2007.

Hall, P.P., 2000: *Urban Future 21: A Global Agenda for Twenty-First Century Cities* (London: Spon).

Hansen, K.T.; Vaa, M., 2002: *Reconsidering Informality: Perspectives from Urban Africa* (Uppsala, Sweden: Nordiska Afrikainstitutet): 240.

Hardoy, J.E.; Satterthwaite, D., 1989: *Squatter Citizen: Life in the Urban Third World* (London: Earthscan Publications Ltd.).

Harvey, D., 2005: *A Brief History of Neoliberalism David Harvey* (New York: Oxford University Press).

Harvey, D., 2001: "Globalization and the 'Spatial Fix'", in: *Geographische Revue,* 2: 23–30.

Healey, P., 2010: *Making Better Places: The Planning Project in the Twenty-First Century* (New York: Palgrave).

Healey, P., 1997: *Collaborative Planning. Shaping Places in Fragmented Societies* (London: MacMillan).

Hodson, M.; Marvin, S., 2009: "Urban Ecological Security: A New Paradigm?", in: *International Journal of Urban and Regional Research,* 33,1: 193–215.

Holling, C., 1973: "Resilience and Stability of Ecological Systems", in: *Annual Review of Ecology and Systematics,* 4,1: 1–23.

Hoselitz, B.F., 1957: "Urbanization and Economic Growth in Asia", in: *Economic Development and Cultural Change,* 6,1: 42–54.

Howorth, C.; O'Keefe, P.; Convery, I., 1998: *Urban Agriculture in Dar es Salaam, Tanzania* (Newcastle: University of Northumbria, Division of Geography and Environmental Management).

Iaquinta, D.L.; Drescher, A.W., 2001: "More than the Spatial Fringe: An Application of the Peri-urban Typology to Planning and Management of Natural Resources", Paper Prepared for the Conference on Rural–Urban Encounters: Managing the Environment of the Peri-urban Interface, Development Planning Unit, University College of London, 9–10 November.

IISD, 2012: Earth Negotiations Bulletin—Doha Climate Change Conference Published by the International Institute for Sustainable Development (IISD). Vol. 12 No. 556. http://www.iisd.ca/climate/cop18/enb/26nov.html.

IPCC, 2001: "Climate Change 2001: Impacts, Adaptation and Vulnerability. Contribution of Working Group II to the Third Assessment Report of the Intergovernmental Panel on Climate Change", in: McCarthy, J.J.; Canziani, O.F.; Leary, N.A.; Dokken, D.J.; White, K.S. (Eds.): (Cambridge: Cambridge University Press). Smit, B. et al (Cap. 18), http://www.grida.no/publications/other/ipcc_tar/.

IPCC, 2007a: "Climate Change 2007: Contribution of Working Group II to the Fourth Assessment Report of the Intergovernmental Panel on Climate Change, 2007", in: Parry, M.L.; Canziani, O.F.; Palutikof, J.P.; van der Linden, P.J.; Hanson, C.E. (Eds.): (Cambridge: Cambridge University Press). (Cap 20), http://www.ipcc.ch/publications_and_data/publications_ipcc_fourth_assessment_report_wg2_report_impacts_adaptation_and_vulnerability.htm.

IPCC, 2007b: "Climate Change 2007: Synthesis Report. Contribution of Working Groups I, II and III to the Fourth Assessment Report of the Intergovernmental Panel on Climate Change" [Core Writing Team, Pachauri, R.K.; Reisinger, A. (Eds.)] (Geneva, Switzerland: IPCC) http://www.ipcc.ch/publications_and_data/ar4/syr/en/contents.html.

IPCC, 2007c: "Climate Change 2007: Impacts, Adaptation and Vulnerability. Summary for Policymakers. Contribution of Working Group II to the Fourth Assessment Report of the Intergovernmental Panel on Climate Change", Summary Approved at the 8th Session of Working Group II of the IPCC, Brussels, April 2007.

IPCC, 2007d: "Impacts, Adaptation and Vulnerability. Contribution of Working Group II to the Fourth Assessment Report of the Intergovernmental Panel on Climate Change. Full Report", in: Parry, M.L.; Canziani, O.F.; Palutikof, J.P.; van der Linden, P.J.; Hanson, C.E. (Eds.): *2007, Climate Change 2007: Impacts, Adaptation and Vulnerability. Contribution of Working Group II to the Fourth Assessment Report of the Intergovernmental Panel on Climate Change* (Cambridge: Cambridge University Press): 982.

IPCC, 2012: *Managing the Risks of Extreme Events and Disasters to Advance Climate Change Adaptation. A Special Report of Working Groups I and II of the Intergovernmental Panel on Climate Change* (Cambridge, UK, and New York, USA: Cambridge University Press).

IPCC, 2014a: "Summary for Policymakers", in: Field, C.B.; Barros, V.R.; Dokken, D.J.; Mach, K.J.; Mastrandrea, M.D.; Bilir, T.E.; Chatterjee, M.; Ebi, K.L.; Estrada, Y.O.; Genova, R.C.; Girma, B.; Kissel, E.S.; Levy, A.N.; MacCracken, S.; Mastrandrea, P.R.; White, L.L. (Eds.): *Climate Change 2014: Impacts, Adaptation, and Vulnerability. Part A: Global and Sectoral Aspects. Contribution of Working Group II to the Fifth Assessment Report of the Intergovernmental Panel on Climate Change* (Cambridge: Cambridge University Press) 1–32.

IPCC, 2014b: *Climate Change 2014: Impacts, Adaptation, and Vulnerability. Part A: Global and Sectoral Aspects. Contribution of Working Group II to the Fifth Assessment Report of the Intergovernmental Panel on Climate Change* [Field, C.B.; Barros, V.R.; Dokken, D.J.; Mach, K.J.; Mastrandrea, M.D.; Bilir, T.E.; Chatterjee, M.; Ebi, K.L.; Estrada, Y.O.; Genova, R.C.; Girma, B.; Kissel, E.S.; Levy, A.N.; MacCracken, S.; Mastrandrea, P.R.; White, L.L. (Eds.)] (Cambridge: Cambridge University Press): 1132.

IPCC, 2014c: *Climate Change 2014: Impacts, Adaptation, and Vulnerability. Part B: Regional Aspects. Contribution of Working Group II to the Fifth Assessment Report of the Intergovernmental Panel on Climate Change* [Barros, V.R., Field, C.B.; Dokken, D.J.; Mastrandrea, M.D.; Mach, K.J.; Bilir, T.E.; Chatterjee, M.; Ebi, K.L.; Estrada, Y.O.; Genova, R.C.; Girma, B.; Kissel, E.S.; Levy, A.N.; MacCracken, S.; Mastrandrea, P.R.; White, L.L. (Eds.)] (Cambridge: Cambridge University Press): 688.

ISD Reporting Services, 2007: "Second International Workshop on Community-Based Adaptation on Climate Change: 24–28 February 2007", in: *Community Based Adaptation to Climate Change Bulletin*, 135, 1, http://www.iisd.ca/ymb/sdban/.

Jacobs, J., 1996: *Edge of Empire: Postcolonialism and the City* (New York: Routledge).

Jamal, V.; Weeks, J., 1993: *Africa Misunderstood: Or Whatever Happened to the Rural-Urban Gap?* (London: Macmillan).

Janssen, M.A.; Ostrom, E., 2006: "Empirically Based, Agent-Based Models", in: *Ecology and Society,* 11,2: 37.

Jones, G.A.; Corbridge, S., 2010: "The Continuing Debate About Urban Bias: The Thesis, Its Critics, Its Influence and Its Implications for Poverty-Reduction Strategies", in: *Progress in Development Studies,* 10,1: 1–18.

Kates, R.W.; Travis, W.R.; Wilbanks, T.J., 2012: "Transformational Adaptation When Incremental Adaptations to Climate Change are Insufficient", in: *Proceedings of the National Academy of Sciences*, 109,19: 7156–7161.

Kebede, A.S.; Nicholls, R.J., 2012: "Exposure and Vulnerability to Climate Extremes: Population and Asset Exposure to Coastal Flooding in Dar es Salaam, Tanzania", in: *Regional Environmental Change,* 12,1, 81–94.

Kironde, J.M.L., 1995: "The Evolution of the Land Use Structure of Dar es Salaam 1890–1990: A Study into the Effects of Land Policy" (Ph.D. Dissertation, University of Nairobi).
Kironde, J.M.L., 2001: *Peri-urban Land Tenure, Planning and Regularisation: Case Study of Dar es Salaam, Tanzania, Harare, Zimbabwe: Study Carried out for the Municipal Development Programme,* Harare, Zimbabwe.
Kironde, J.M.L., 2006: "The Regulatory Framework, Unplanned Development and Urban Poverty: Findings from Dar es Salaam, Tanzania", in: *Land Use Policy,* 23,4: 460–472.
Kiunsi, R., 2013. "The Constraints on Climate Change Adaptation in a City with a Large Development Deficit: The Case Of Dar es Salaam", in: *Environment and Urbanization*, 25,2: 321–337.
Kombe, W.J., 2005: "Land Use Dynamics in Peri-urban Areas and Their Implications on the Urban Growth and Form: The Case of Dar es Salaam, Tanzania", in: *Habitat International*, 29,1: 113–135.
Kombe, W.J.; Kreibach, V., 2000: *Informal Land Management in Tanzania*. Dortmund: SPRING Research Series no. 29.
Klein, R.J.T.; Nicholls, R.J.; Thomalla F., 2003: "The Resilience of Coastal Megacities to Weather Related Hazards: A Review", in: Kreimer, A.; Arnold, M.; Carlin, A. (Eds.): *Building Safer Cities: The Future of Disaster Risk, Disaster Risk Management Series No. 3* (Washington, D.C.: World Bank): 111–137.
Kyessi, S.A.; Misigaro, A.; Shoo J., 2009: "Formalisation of Land Property Rigths in Unplanned Settlements: Case of Dar es Salaam, Tanzania", in: *TPAT*, 6.
Latouche, S., 1993: *Il pianeta dei naufraghi. Saggio sul sottosviluppo* (Torino: Bollati Boringhieri) (ed. or. 1991: La planète des naufragés. Paris, La Découverte).
Latouche, S., 1997: *L'altra Africa. Tra dono e mercato* (Torino: Bollati Boringhieri).
Latouche, S., 2005: *L'invenzione dell'economia* (Torino: Bollati Boringhieri).
Lefebvre, H., 1991: *The Production of Space* (Oxford: Blackwell).
Lee-Smith, D.; Stren, R., 1991: "New Perspectives on African Urban Management", in: *Environment and Urbanization*, 3,l: 23–36.
Leitmann, J., 1999: *Sustaining Cities. Environmental Planning and Management in Urban Design* (New York: McGraw-Hill).
Lerise, F., 2000: "Centralised Spatial Planning Practice and Land Development Realities in Rural Tanzania", in: *Habitat International*, 24,2: 185–200.
Leslie, J.A., 1963: *A Survey of Dar es Salaam* (London: Oxford University Press).
Lewis, W.A., 1954: "Economic Development with Unlimited Supplies of Labour", in: *Manchester School of Economic and Social Studies*, 22: 139–191.
Lindell, I., 2010: *Africa's Informal Workers: Collective Agency, Alliances and Transnational Organizing in Urban Africa* (London and Uppsala: Zed Books and The Nordic Africa Institute).
Lindell, I., 2010a: "Between Exit and Voice: Informality and the Spaces of Popular Agency", in: *Special Issue of African Studies Quarterly,* 11,2/3: 1–11.
Lipton, M., 1977: *Why Poor People Stay Poor: A Study of Urban Bias in World Development* (London: Temple Smith).
Lourenço-Lindell, L., 2002: "Walking the Tight Rope: Informal Livelihoods and Social Networks in a West African" (Doctoral Thesis, University of Uppsala, Stockholm, Acta 9, 3 June 2002).
Luers, A.L., 2005: "The Surface of Vulnerability: An Analytical Framework for Examining Environmental Change", in: *Global Environmental Change*, 15: 214–223.
Lupala, A., 2002a: "Peri-Urban Land Management in Rapidly Growing Cities, The Case of Dar es Salaam" (Ph.D. Dissertation, University of Dortmund).
Lupala, A., 2002b: "The Dynamics of Peri-urban Growth in Dar es Salaam", in: *Spring Research Series,* 31.
MacGregor, S., 1995: "Planning Change: Not an End but a Beginning", in: Eichler, M. (Ed.): *Change of Plans: Towards a Non-sexist Sustainable City* (Toronto: Garamond Press): 151–167.

Mabin, A.; Butcher, S.; Bloch, R., 2013: "Peripheries, Suburbanisms and Change in Sub-Saharan African Cities", in: *Social Dynamics*, 39,2: 167–190.

Malele, B.F., 2009: "The Contribution of Ineffective Urban Planning Practices to Disaster and Disaster Risks Accumulation in Urban Areas: The Case of Former Kunduchi Quarry Site in Dar es Salaam, Tanzania", in: *JÀMBÁ: Journal of Disaster Risk Studies*, 2,1: 28–53.

Marcuse, P., 2004: "Said's Orientalism: A Vital Contribution Today", in: *Antipode,* 36: 809–817.

Marshall, N.A.; Marshall, P.A.; Tamelander, J.; Obura, D.; Malleret-King, D.; Cinner, J.E., 2010: A Framework for Social Adaptation to Climate Change; Sustaining Tropical Coastal Communities and Industries. Gland, Switzerland, IUCN, 36 pp.

Masanja, A.L., 2002: "Rural-Urban Dynamics: Modelling and Predicting Peri-urban and Use Changes in Dar es Salaam city, Tanzania", Proceedings IHDP: Urbanization and the Transition to Sustainability, Bonn.

Massey, D., 2005: *For Space* (Sage: London).

Matthews, T., 2011: "Climate Change Adaptation in Urban Systems: Strategies for Planning Regimes", Research Paper 32, Urban Research Program, Griffith University.

Mattingly, M., 1999: "Institutional Structures and Processes for Environmental Planning and Management of the Peri-urban Interface", Paper Produced for the Research Project on Strategic Environmental Planning and Management for the Peri-urban Interface. Development Planning Unit, University College London.

Mattingly, M., 2009: "Making Land Work for the Losers. Policy Responses to the Urbanization of Rural Livelihoods", in: *International Development Planning Review,* 31,1: 37–64.

Mattingly, M., 2006: "Synthesis of Peri-urban Interface Knowledge Natural Resources Systems Programme", Final Technical Report 1 Dfid Project Number R8491.

Maxwell, D., 1999: "The Political Economy of Urban Food Security in Sub-Saharan Africa", in: *World Development,* 27,11: 1939–1953.

Mbembe, A., 2001: *On the Postcolony. Studies on the History of Society and Culture* (Berkeley: University of California Press).

Mbembe, A., 2005: *Postcolonialismo* (Roma: Meltemi Editore srl).

Mbembe, A.; Nuttall, S., 2004: "Writing the World from an African Metropolis", in: *Public Culture,* 16,3: 347–372.

Mbiba, B.; Huchzermeyer, M., 2002: "Contentious Development: Peri-urban Studies in Sub-Saharan Africa", in: *Progress in Development Studies,* 2,2: 113–131.

Mbonile, M.; Kivelia, J., 2008: "Population, Environment and Development in Kinondoni District, Dar es Salaam", in: *Geographical Journal,* 174,2: 169–175.

McAuslan, P., 1985: *Urban Land and Shelter for the Poor* (London: Earthscan).

McCarney, P.; Halfani, M.; Rodriquez, A., 1995: "Towards an Understanding of Governance: The Emergence of an Idea and Its Implications for Urban Research in Developing Countries", in: Stren, R.; Kjellberg, B. (Eds.): *Perspectives on the City University of Toronto* (Toronto: Centre for Urban and Community Studies): 91–142.

McGranahan, G.; Satterthwaite, D., 2000: "Forthcoming: Environmental Health or Ecological Sustainability: Reconciling the Brown and Green Agendas in Urban Development", in: Pugh, C. (Ed.): *Sustainable Cities in Developing Countries* (London: Earthscan).

McGregor, D.; Simon, D.; Thompson, D. (Eds.), 2006: *The Peri-urban Interface: Approaches to Sustainable Natural and Human Resource Use* (London/Stirling, VA: Earthscan).

Meagher, K., 2010: "The Tangled Web of Associational Life: Urban Governance and the Politics of Popular Livelihoods in Nigeria", in: *Urban Forum,* 21,3: 299–313.

Meeus, S.; Gulinck, H., 2008: "Semi-urban Areas in Landscape Research: A Review", in: *Living Reviews in Landscape Research*, 2: 1–45.

Muller, M., 2007: "Adapting to Climate Change: Water Management for Urban Resilience", in: *Environment and Urbanisation,* 19,1: 99–113.

Murray, M.J.; Myers, G.A., 2006: *Cities in Contemporary Africa* (New York: Palgrave Macmillan).

Mwamfupe, D., 2007: "Urban Expansion in Dar es Salaam City in Tanzania: A Blight or Blessing for Peri-urban Livelihoods?", in: *Afrika Tamulmanyok*, 2: 1–14.
Myers, G.A., 2011: *African Cities. Alternative Visions of Urban Theory and Practice* (London and New York: Zed Books).
Myers, G.A., 2003: "Colonial and Postcolonial Modernities in Two African Cities", in: *Canadian Journal of African Studies/Revue Canadienne des Études Africaines,* 37,2/3: 328–357.
Myers, G.A., 2005: *Disposable Cities: Garbage, Governance and Sustainable Development in Urban Africa* (Aldershot, Hants, England: Ashgate).
Nelson, D.R.; Adger, W.N.; Brown, K., 2007: "Adaptation to Environmental Change: Contributions of a Resilience Framework", in: *Annual Review of Environmental and Resources*, 32: 395–419.
Nelson, S.C., 2007: Farming on the Fringes: Changes in Agriculture, Land Use and Livelihoods in Peri-Urban Dar es Salaam, Tanzania. *Honors Projects*. Paper 10.
Nicholls, R.J.; Klein, R.J.T.; Tol, R.S.J., 2007: "Managing Coastal Vulnerability and Climate Change: A National to Global Perspective", in: McFadden, et al. (Eds.) *Managing Coastal Vulnerability* (Oxford: Elsevier): 223–241.
Ngau, P., 2013: For Town and Country: A New Approach to Urban Planning in Kenya, Africa Research Institute. Policy Voices Series http://www.africaresearchinstitute.org/publications/policy-voices/urban-planning-in-kenya/.
O'Brien, K.; Eriksen, S.; Nygaar, L.P.; Schjolden A., 2007: "Why Different Interpretations of Vulnerability Matter in Climate Change Discourses", in: *Climate Policy,* 2007,7: 73–88.
Odendaal, N., 2012: "Reality Check: Planning Education in the African Urban Century", in: *Cities,* 29: 174–182.
OECD, 2007: Working Party on Global and Structural Policies. Literature Review on Climate Change Impacts on Urban City Centres. Hunt, A., Watkiss, P. Ref. ENV/EPOC/GSP(2007) 10/FINAL.
Ostrom, E., 1990: *Governing the Commons: The Evolution of Institutions for Collective Action* (Cambridge: Cambridge University Press).
Pahl-Wostl, C., 2009: "A Conceptual Framework for Analysing Adaptive Capacity and Multi-level Learning Processes in Resource Governance Regimes", in: *Global Environmental Change*, 18: 354–365.
Parnell, S.; Pieterse, E.A., 2014: *Africa's Urban Revolution* (London, New York: Zed Books).
Parnell, S.; Simon, D., 2014: "National Urbanization and Urban Strategies: Necessary but Absent Policy Instruments in Africa", in: Parnell, S.; Pieterse, E. (Eds.): *Africa's Urban Revolution* (London and New York: Zed Books): 237–256.
Pelling, M., 2011: *Adaptation to Climate Change: From Resilience to Transformation* (Abingdon, Oxon, England; New York: Routledge).
Pieterse, E., 2008: *City Futures: Confronting the Crisis of Urban Development* (London: Zed Books).
Pieterse, E., 2009: *Exploratory Notes on African Urbanism.*
Pieterse, E., 2010: "Cityness and African Urban Development", in: *Urban Forum,* 21,3: 205–219.
Polese, M., 2005: "Cities and National Economic Growth: A Reappraisal", in: *Urban Studies* , 42,8: 1429–1451.
Polèse, M., 2010: The Resilient City: On the Determinants of Successful Urban Economies. INRS-UCS, Montréal, 2010.03, 32 pp.
Polsky, C., et al., 2007: "Building Comparable Global Change Vulnerability Assessments: The Vulnerability Scoping Diagram", in: *Global Environmental Change,* 17,3–4: 472–485.
Rakodi, C., 1997: *The Urban Challenge in Africa: Growth and Management of Its Large Cities* (Tokyo, New York: United Nations University).
Rakodi, C., 1998: *Review of the Poverty Relevance of the Peri-urban Interface Production System Research.* Report for the DFID Natural Resources Systems Research Programme (PD 70/7E0091), second draft.

Rakodi, C.; Lloyd-Jones, T., 2002: *Urban Livelihoods: A People-Centred Approach to Reducing Poverty* (London: Earthscan).
Ricci, L.; Demurtas, P.; Macchi, S.; Cerbara, L., 2012a: "Investigating the Livelihoods of the Population Dependent on Natural Resources and their Concerns Regarding Climate Change", Working Paper, ACCDAR Project WP 1 Activity 1.1. http://www.planning4adaptation.eu/Docs/papers/08_Working_Paper_Activity_1.1.pdf.
Ricci, L., 2012b: "Peri-urban Livelihood and Adaptive Capacity: Urban Development in Dar Es Salaam", in: *Consilience: The Journal of Sustainable Development,* 7,1, 46–63.
Robards, M., et al., 2011: "The Importance of Social Drivers in the Resilient Provision of Ecosystem Services", in: *Global Environmental Change.*
Robards, M.; Schoon, M.; Meek, C.; Engle, N., 2011: "The Importance of Social Drivers in the Resilient Provision of Ecosystem Services", in: *Global Environmental Change,* 21,2: 522–529.
Robinson, J., 2002: "Global and World Cities: A View from off the Map", in: *International Journal of Urban and Regional Research,* 26: 531–554.
Robinson, J., 2003: "Postcolonialising Geography: Tactics and Pitfalls", in: *Singapore Journal of Tropical Geography,* 24: 273–289.
Robinson, J., 2006: *Ordinary Cities: Between Modernity and Development* (London and New York: Routledge).
Robinson, J., 2010: "Cities in a World of Cities: The Comparative Gesture", in: *International Journal of Urban and Regional Research,* 35,1: 1–23.
Romero Lankao, P.; Qin, H., 2011: "Conceptualizing Urban Vulnerabilità to Global Cliamte and Environmental Change", in: *Current Opinion in Environmental Sustainability.*
Romero-Lankao, P., 2007: "Are We Missing the Point? Particularities of Urbanization, Sustainability and Carbon Emission in Latin America Cities", in: *Environment and Urbanization,* 19,1: 159–175.
Romero-Lankao, P., 2008: "Urban Areas and Climate Change: Review of Current Issues and Trends", in: *Issues Paper for the 2011 Global Report on Human Settlements.*
Roy, A., 2002: *City Requiem, Calcutta: Gender and the Politics of Poverty (Globalization and Community)* (Minneapolis: University of Minnesota Press).
Roy, A., 2005: "Urban Informality", in: *Journal of the American Planning Associarion,* 71,2: 147–159.
Roy, A., 2009: "The 21st-Century Metropolis: New Geographies of Theory", in: *Regional Studies,* 43,6: 819–830.
Roy, A.; Al Sayyad, N., 2004: *Urban Informality: Transnational perspectives fom the Middle East, South Asia and Latin America* (Lanham, MD: Lexington Books).
Said, W.E., 1978: *Orientalism* (London: Penguin).
Said, W.E., 1985: "Orientalism Reconsidered", in: *Cultural Critique,* 1: 89–107.
Sandercock, L., 2003: "Toward Cosmopolis: Utopia as Construction Site", in: Campbell, S.; Fainstein, S.S. (Eds.): *Readings in Planning Theory*, 2nd edn. (Oxford: Blackwell): 401–410.
Sanderson, D., 2000: "Cities, Disasters and Livelihoods", in: *Environment and Urbanization*, 12, 93–102.
Sappa, G.; Trotta, A.; Vitale, S., 2014: "Climate Change Impacts on Groundwater Active Recharge in Coastal Plain of Dar es Salaam (Tanzania)", in: IAEG (Ed.): *Engineering Geology for Society and Territory*, vol. 1 (New York: Springer International Publishing): 177–180.
Sassen, S., 1991: *The Global City: New York, London, Tokyo* (Princeton, NJ: Princeton University Press).
Satterthwaite, D., 2008: "'Cities' Contribution to Global Warming: Notes on Allocation of Greenhouse Gas Emission", in: *Environment and Urbanizaztion,* 20,2: 539–549.
Satterthwaite, D.H.; Pelling, M.; Reid, A.; Romero-Lankao, P., 2007: *Building Climate Change Resilience in Urban Areas and among Urban Populations in Low- and Middle-income Nations.* IIED Research report commissioned By the Rockfeller Foundation.
Sawio, C.J., 1994: *Urban Agriculture and Sustainable Dar es Salaam Project, Tanzania.* UNCHS-IDRC Project Coordinator.

Sen, A., 1999: *Development as Freedom* (New York: Knopf).

Sennett, R., 1990: *The Conscience of the Eye—The Design and Social Life of Cities* (London: Faber and Faber).

Sheuya, S.A., 2010: *Informal Settlements and Finance in Dar es Salaam, Tanzania* (Nairobi: UN-HABITAT).

Simon, D., 2003: "Regional Development-Environment Discourses, Policies and Practices in Post-Apartheid Southern Africa", in: Grant, J.; Søderbaum, F. (Eds.): *The New Regionalism in Africa* (Aldershot: Ashgate): 67–89.

Simon, D., 2008: "Urban Environments: Issues on the Peri-Urban Fringe", in: *Annual Review of Environment and Resources,* 33: 167–185.

Simon, D., 2013: "Climate and Environmental Change and the Potential for Greening African Cities'", in: *Local Economy,* 28, 203–217.

Simon, D.; McGregor, D.; Nsiah-Gyabaah, K., 2004: "The Changing Urban-Rural Interface of African Cities: Definitional Issues and an Application to Kumasi, Ghana", in: *Environment and Urbanization,* 16: 235–247.

Simone, A., 2004a: *For the City Yet to Come: Changing African Life in Four Cities* (Durham & London: Duke University Press): x+297.

Simone, A., 2004b: "People as Infrastructure: Interectin Fragments in Joannesburg", in: *Public Culture,* 16,3: 407–429.

Simone, A., 2010: *Citi Life from Jakarta to Dakar: Movements at Crossroads* (London: Routledge).

Simone, A.; Abouhani, A., 2005: *Urban Africa: Changing Contours of Survival in the City* (Dakar: Cordesia).

Sliuzas, R., 2004: *Managing Informal Settlements: A Study Using Geo-information in Dar es Salaam, Tanzania.* Enschede, ITC, ITC Dissertation 112.

Smit, B.; Pilifosova, O.; Burton, I.; Challenger, B.; Huq, S.; Klein, R.J.T.; Yohe, G.; Adger, N.; Downing, T.; Harvey, E.; Kane, S.; Parry, M.; Skinner, M.; Smith, J.; Wandel, J., 2001: *Adaptation to Climate Change in the Context of Sustainable Development (s.d.).*

Smit, B.; Wandel, J., 2006: "Adaptation, Adaptive Capacity and Vulnerability", in: *Global Environmental Change*, 16,3: 282–292.

Smucker, T.; Wisner, B.; Mascarenhas, A.; Munishi, P.K.; Wangui, E.E.; Sinha, G.; Weiner, D.; Bwenge, C.; Lovell, E., 2015: "Differentiated Livelihoods, Local Institutions, and the Adaptation Imperative: Assessing Climate Change Adaptation Policy in Tanzania", in: *Geoforum* 59: 39–50.

Sovani, N.V., 1964: "The Analysis of Over-Urbanization", in: *Economic Development and Cultural Change* (University of Chicago Press), 12,2: 113–122.

Stern, N., 2006: *The Economics of Climate Change: The Stern Review* (Cambridge, UK: Cambridge University Press).

Stren, R.; Eyoh, D., 2007: "Decentralization and Urban Development in West Africa: An introduction", in: Eyoh, D.; Stren, R. (Eds.): *Decentralization and the Politics of Urban Development in West Africa* (Washington, DC: Woodrow Wilson International Center for Scholars).

Sutton, 1970: Dar es Salaam. City, Port and Region, edited da J.E.G. Sutton Tanzania Notes and Records, n.71 Pubblicato da Tanzania Society, 1970.

Tacoli, C., 1998a: "Beyond the Rural-Urban Divide", in: *Environment & Urbanization*, 10,1: 3–4.

Tacoli, C., 1998b: "Rural–Urban Interactions; a Guide to the Literature", in: *Environment & Urbanization,*10,1: 147–166.

Tanner, T.; Mitchell, T.; Polack, E.; Guenther, B., 2009: *Urban Governance for Adaptation: Assessing Climate Change Resilience in Ten Asian Cities.* IDS Working Paper 315 (Brighton (UK): Institute of Development Studies).

Tarafdar, A.K.; Bjonness, H.C., 2011: "Environmental Premises in Planning for Sustainability at Local Level in Large Southern Cities: A Case Study in Kolkata, India and Use of the PRETAB Planning Process Model", in: *International Journal of Sustainable Development & World Ecology,* 17,1: 24–38.

Todes, A., 2011: "Reinventing Planning: Critical Reflections", in: *Urban Forum,* 22,2: 115–133.

Tostensen, A.; Tvedten, I.; Vaa, M., 2001: *Associational Life in African Cities: Popular Responses to the Urban Crisis* (Uppsala: Nordic Africa Institute).

Tripp, A.M., 1997: *Changing the Rules: The Politics of Liberalization and the Urban Informal Economy in Tanzania* (Berkeley/Los Angeles: University of California Press).

Turner, J.C., 1968: "Housing Priorities, Settlement Patterns, and Urban Development in Modernizing Countries", in: *AIP Journal* (November), xxxiv,6: 354–363.

Turok, I., 2014: "Linking Urbanization and Development in Africa's Economic Revival", in: Parnell, S.; Pieterse, E. (Eds.): *Africa's Urban Revolution* (London and New York: Zed Books): 60–81.

UN, 2005: *Una casa nella città. Progetto del Millennio delle Nazioni Unite 2005.* Task Force on Improving of lives of Slums Dwellers. Edizione italiana DGCS, 2005.

UN, 2008: *World Urbanization Prospects: The 2007 Revision,* CD-ROM Edn., data in digital form (POP/DB/WUP/Rev.2007) (New York: United Nations, Department of Economic and Social Affairs, Population Division).

UN-DESA, 2014: *World Urbanization Prospects: The 2014 Revision* (New York: Department of Economic and Social Affairs, Population Division).

UN Habitat, 2008: *State of the African Cities Report 2008: A Framework for Addressing Urban Challenges in Africa* (Nairobi: UN-HABITAT).

UN-Habitat, 2011: *Cities and Climate Change: Global Report on Human Settlements. United Nations Human Settlements Programme* (London, Washington DC: Earthscan).

UN-HABITAT, 2010: *Citywide Action Plan for Upgrading Unplanned and Unserviced Settlements in Dar es Salaam* (Nairobi: United Nations Human Settlements Programme (UN-HABITAT)).

United Republic of Tanzania—URT, 1997: *National Land Policy,* 2nd Edn. (Dar es Salaam: Ministry of Lands and Human Settlements Development).

United Republic of Tanzania—URT, 1999: *The Land Act, 1999 (Act No. 4 of 1999)* (Dar es Salaam: Ministry of Lands and Human Settlements Development. Government Printers).

United Republic of Tanzania—URT, 2005: *National Strategy for Growth and Reduction of Poverty (NSGRP)* (Dar es Salaam: Vice President's Office).

United Republic of Tanzania—UTR, 2007: *The National Land Use Planning Act, 2007* (Dar es Salaam: Government Printer).

United Republic of Tanzania—URT, 2011: *The Dar es Salaam City Environment Outlook 2011, Draft* (Dar es Salaam: Vice President's Office, Division of Environment).

Varshney, A., 1993: *Beyond Urban Bias* (New York: Routledge).

Walker, B.H.; Meyers, J.A., 2004: "Thresholds in Ecological and Social-Ecological Systems: A Developing Data Base", in: *Ecological and Society*, 9,2.

Walker, B.; Gunderson, L.; Kinzig, A.; Folke, C.; Carpenter, S.; Schultz, L., 2006: "A Handful of Heuristic and Some Propositions for Understanding Resilience in Socio-Ecological Systems", in: *Ecological and Society*, 11.

Watts, M., 2005: "Baudelaire over Berea, Simmel over Sandton?", in: *Public Culture*, 17: 181–192.

Watts, M.J.; Bohle, H.-G., 1993: "The Space of Vulnerability: The Causal Structure of Hunger and Famine", in: *Progress in Human Geography,* 17: 43–67.

Annex I
Questionnaires

Questionnaire A—Household Survey ***(Dodoso ngazi ya familia)***

Personal data *(maelezo binafsi)*
Name *(Jina)*
Gender *(Jinsia)*
Age *(umri)*
Job *(kazi)*
Education *(elimu)*
Contact *(mawasiliano)*
Place *(mahali)*
Date (*tarehe*)

1. Have you been here before? How long have you lived here? *(Umewahi kuwa maeneo haya kabla? Umeisha hapa kwa muda gani?)*

2. Where are you from? *(Umehamia kutoka wapi?)*

a. The same district (Kinondoni) *(ndani ya wilaya ya kinondoni)*
b. The same region (Dar Es Salaam) *(ndani ya mkoa wa Dar es Salaam)*
c. Other regions *(mikoa mingine?)*
d. Other *(kuingineko)*

3. Why did you come here? Did you come here alone or with your family? *(Kwa nini ulihamia hapa? Ulikuja peke yako au pamoja na familia yako?)*

4. What is the size of your household? *(Familia yako ina watu wangapi?)*

L. Ricci, *Reinterpreting Sub-Saharan Cities through the Concept of Adaptive Capacity*, SpringerBriefs in Environment, Security, Development and Peace 26, DOI 10.1007/978-3-319-27126-2

5. Did you buy this piece of land? (If yes, how much did it cost?) Do you have the ownership of this land? How big is it? *(Ulinunua hili shamba? Kama jibu ni ndiyo liligharimu kiasi gani?)*

6. Do you pay any tax and/or fee for land tenure or facilities? (If yes, specify how much) *(Unalipa kodi ya ardhi au huduma zingine za jamii? Kama jibu ni ndiyo ainisha kiasi unacholipa)*

a. Water *(maji)*
b. Waste *(uchafu)*
c. Electricity *(umeme)*
d. Land/house *(ardhi/nyumba)*
e. Other *(nyinginezo)*

7. Describe your main activities *(Elezea shughuli kuu ufanyazo)*

a. Agriculture (specify crop) *(kilimo na aina ya mazao)*
b. Livestock (specify animal) *(ufugaji na aina ya mifugo)*
c. Charcoal making *(uchomaji mkaa)*
d. Other *(nyinginezo)*

8. How often do you go to the city? *(Ni mara ngapi unaenda mjini?)*

a. More or less than once a day *(zaidi au chini ya mara moja kwa siku)*
b. More or less than once a week *(zaidi au chini ya mara moja kwa wiki)*
c. More or less than twice a month *(zaidi au chini ya mara mbili kwa mwezi)*
d. Rarely *(nadrila)*

9. Which transport do you usually use? *(kwa kawaida unatumia usafiri gani?)*

a. Foot *(miguu)*
b. Daladala (or other bus/minibus) *(Daladala)*
c. Car *(gari)*
d. Bicycle *(baiskeli)*
e. Motorcycle (pikipiki)
f. Other *(njia nyingine kama zipo)*

10. How and where you restock water for domestic use (specify other uses)? How many liters per day you need? *(Ni vipi na wapi mnahifadhi maji kwa matumizi ya nyumbani (ainisha matumizi mengine)? Unahitaji lita ngapi za maji kwa siku?)*

a. Tank *(tenki)*
b. Other containers *(vyombo vingine)*

11. Do you have electricity in your house? How long you use it per day? *(nyumba yako ina umeme? Unatumia masaa mangapi kwa siku?)*

12. How and where you collect solid waste (a–c) and wastewater (d–f)? *(mnakusanyaje taka ngumu (a–c) na maji taka (d–f) na zinatupwa wapi?)*

a. Public or private collection (specify the frequency) *(serikali au kampuni binafsi, mara ngapi?)*
b. Individual collection *(binafsi)*
c. Other *(nyinginezo)*
d. public or private collection (specify the frequency) *(serikali au kampuni binafsi, mara ngapi?)*
e. Individual collection *(binafsi)*
f. Other *(nyinginezo)*

13. What is estimated amount of solid waste generated per day? *(mnazalisha kiasi gani cha taka ngumu kwa siku?)*

a. 1 bucket *(ndoo 1)*
b. 1–3 buckets *(ndoo 1–3)*
c. More than 3 buckets *(zaidi ya ndoo 3)*

14. Do you practice any treatment of solid wastes (a–c) and wastewater (d–f)? *(mna mfumo wowote maalumu wa usimamizi wa taka ngumu (a–c) na maji taka (d–f)?)*

a. Recycling (specify how) *(kurudisha kiwandani)*
b. Composting *(kutengeneza mbolea)*
c. Other *(njia nyinginezo)*
d. Reuse/Recycling (specify how) *(matumizi mengine au kurudisha kiwandani)*
e. Composting *(kutengeneza mbolea)*
f. Other *(njia nyinginezo)*

15. How much is your income now? *(unadhani kipato chako kwa sasa ni kiasi gani?)*

16. Would you move to the city centre or countryside in the future? Why? *(una mpango wowote wa kuhamia mjini au pembezoni mwa nchi? Kwa nini?)*

17. Have you observed environmental changes during the last year? *(katika miaka iliyopita, kuna mabadiliko yoyote ya kimazingira umeyaona?)*

a. Water availability *(upatikanaji wa maji)*
b. Soil fertility *(rutuba kwenye udongo)*
c. Soil aridity *(ukavu kwenye udongo)*
d. Humidity *(unyevunyevu kwenye hewa)*
e. Rain pattern *(mzunguko wa mvua)*
f. Other *(mengineyo)*

18. Have you changed your activities or attitudes because of the changes above? *(Je umebadilisha shughuli au mwenendo wako kwa sababu ya mabadiliko yaliyotajwa hapo juu?)*

a. Change in crop system *(mabadiliko ya mfumo wa mazao)*
b. Change in livestock *(mabadiliko katika ufugaji)*
c. Change on house structure *(mabadiliko kwenye ujenzi wa nyumba)*

19. Do you have some short and/or long-term strategies or activities to cope with environmental stresses (drought, flooding, …) ? (specify what and how long) *(Mna mikakati yoyote ya muda mfupi au mrefu au shughuli za kukabiliana na madhara ya kimazingira (ukame, mafuliko n.k)? (ainisha))*

20. What are the reasons of these environmental problems? *(Nini hasa sababu ya haya matatizo ya mazingira?)*

a. Climatic variation *(mabadiliko ya tabia ya nchi)*
b. Human land use *(matumizi ya ardhi)*
c. Inadequate environmental management from institutions *(mapungufu katika usimamizi wa mazingira kutoka taasisi husika)*
d. Other *(nyinginezo)*

21. Notes

Membership of group, network, access to the wider institutional society *(makundi ya uanachama, mtandao, namna ya upatikanaji wa huduma kutoka taasisi mbalimbali)*

Education availability *(upatikanaji wa elimu)*

Health availability *(upatikanaji wa huduma ya jamii)*

Questionnaire B—Ward Leaders Survey *(Dodoso kwa viongozi wa kata)*

Personal data *(maelezo binafsi)*
Name *(Jina)*
Gender *(Jinsia)*
Age *(umri)*
Job (position) *(kazi-cheo)*
Education *(elimu)*
Contact *(mawasiliano)*
Place *(mahali)*
Date *(tarehe)*

1. How long have you been working here? *(umefanya kazi hapa kwa muda gani?)*

2. What about your competences and responsibility? Describe your main activities and relation with District and Subwards. *(Kwa uzoefu ulionao, eleza majukumu yako na uhusiano kati ya wilaya na mitaa/vitongoji)*

a. Physical boundaries *(mipaka)*
b. Natural resources management *(usimamizi wa rasilimali)*
c. Land use planning *(mpango wa matumizi bora ya ardhi)*
d. Development planning *(mpango wa maendeleo)*
e. Other *(mengineyo)*

3. Which are the planning approaches and tools at regional and local level in the following sectors? (focused on PU areas) *(Ni mbinu na njia gani katika ngazi ya mkoa na mitaa hutumika kufanikisha mipango katika sekta zifuatazo?)*

a. Environmental risk management *(usimamizi wa hatari za mazingira)*
b. Strategies for natural resources management (water, soil, ...) *(mikakati katika usimamizi wa rasilimali mfano maji na udongo)*
c. Land use planning *(mpango wa matumizi bora ya ardhi)*
d. Facilities supply *(utoaji huduma za jamii)*
e. Urban and/or rural development *(maendeleo ya mjini na vijijini)*
f. Other *(nyinginezo)*

4. Do you consider peri-urban dynamics in planning? Do you have specific policies and strategies for peri-urban areas? What? Are there examples of planning in PU areas? What about results? *(Je mabadiliko ya maeneo ya nje ya mji yanazingatiwa wakati wa mipango? Mnazo sera na mikakati mahususi kwa maeneo ya nje ya mji? Ni zipi? Kuna mifano ya mpango wowote katika maeneo ya nje ya mji? Vipi matokeo yake?)*

5. What you do for sustainable resource management? What about soil, water? *(mnafanya nini kuhakikisha usimamizi endelevu wa rasilimali? Mfano udongo na maji?)*

6. What about donors: names, role/tasks, programs and approaches? *(Vipi kuhusu wahisani: majina, wajibu/kazi, mipango na mbinu wanazotumia?)*

7. What about fiscal systems in PU areas? *(Vipi kuhusu mfumo wa kifedha, mfano kodi na mengineyo kwa maeneo ya nje ya mji?)*

8. What about land tenure in PU areas? *(vipi kuhusu mfumo wa ardhi katika maeneo ya nje ya mji?)*

9. What about facilities supply in PU areas? *(vipi kuhusu utolewaji wa huduma za jamii katika maeneo ya nje ya mji?)*

a. Transport system *(mfumo wa usafiri)*
b. Electricity *(umeme)*
c. Water *(maji)*
d. Waste management *(usimamizi na uzoaji taka)*
e. Education *(elimu)*
f. Health *(afya)*
g. Other *(nyinginezo)*

10. Have you observed changes in PU (U and R) development in recent years? What? *(kuna mabadiliko yoyote yameonekana katika maendeleo ya maeneo ya nje ya mji kwa miaka ya nyuma?)*

11. According to your experiences, what are the main relations between PU-U and PU-R areas? Are these relations changed in the last period? *(kwa maoni yako binafsi, kuna uhusiano gani kati ya maeneo ya nje ya mji na mjini/vijijini? kuna mabadiliko kwenye haya mahusiano ukilinganisha na miaka iliyopita?)*

a. Flows of goods *(upatikanaji wa bidhaa)*
b. People *(watu)*
c. Other *(mengineyo)*

12. What is estimated amount of solid waste generated per day? *(Ni kiasi gani cha taka ngumu huzalishwa kwa siku?)*

a. 1 bucket *(ndoo 1)*
b. 1–3 buckets *(ndoo 1–3)*
c. More than 3 buckets *(zaidi ya ndoo 3)*

13. Do you know about networks and connections, relations of trust and mutual support, formal and informal groups in PU areas? How they contribute in decision-making?

14. Have you observed environmental change during the last years? *(Kuna mabadiliko yoyote ya kimazingira umeyaona katika kipindi cha miaka iliyopita?)*

a. Water availability *(upatikanaji wa maji)*
b. Soil fertility *(rutuba kwenye udongo)*
c. Soil aridity *(ukavu kwenye udongo)*

d. Humidity *(unyevunyevu kwenye hewa)*
e. Rain pattern *(mzunguko wa mvua)*
f. Other *(mengineto)*

15. Have you evaluated the impact of these changes in PU areas? *(Kuna tathimini yoyote imefanyika kuhusu madhara ya mabadiliko ya kimazingira katika maeneo ya nje ya mji?)*

16. Do you implement some strategies to cope with these changes and to reduce PU vulnerability? *(Kuna utekelezaji wa mikakati yoyote ili kukabiliana na hayo mabadiliko na pia kupunguza uwezekano wa kuathirika zaidi kwa maeneo ya nje ya mji?)*

17. Have you observed the autonomous adaptation to these changes? For example changes in PU activities *(umeshaona namna yoyote ya asili katika kukabiliana na hayo mabadiliko? Mfano mabadiliko ya shughuli za maeneo ya nje ya mji)*

a. Change in crop system *(mabadiliko ya mfumo wa mazao)*
b. Change in livestock *(mabadiliko katika ufugaji)*
c. Change on house structure *(mabadiliko katika ujenzi wa nyumba)*

18. Do you have any information or baseline study on PU dwellers composition and livelihoods? *(Mna takwimu au tafiti zozote kuhusu watu na maisha katika maeneo ya nje ya mji)*

a. Employment *(ajira)*
b. Population *(idadi ya watu)*
c. Type and house dimension *(aina na ukubwa wa nyumba)*
d. Other *(nyinginezo)*

19. What are the reasons of these environmental problems? *(Nini sababu hasa ya haya matatizo ya kimazingira?)*

a. Climatic variation *(mabadiliko ya tabia ya nchi)*
b. Human land use *(Matumizi ya ardhi)*
c. Inadequate environmental management *(mapungufu katika usimamizi wa mazingira)*
d. Other *(nyinginezo)*

20. Notes

Have you heard about the National Adaptation Programme of Action (NAPA)? *(umeshasikia kuhusu mpango wa kitaifa wa kukabiliana na mabadiliko ya tabia nchi?)*

How do you implement the NAPA? *(Kama umesikia, mnautekeleza vipi)*

Questionnaire C—District Leaders' Survey

Personal data
Name
Gender
Age
Job (position)
Education
Contact
Place
Date

1. How long have you been working here?

2. What about your competences and responsibility? Describe your main activities and relations with wards, subwards, the region and the state.

a. Physical boundaries
b. Natural resources management
c. Land use planning
d. Development planning
e. Other

3. Which are the planning approaches and tools at regional and local level in the following sectors? (focused on PU areas)

a. Environmental risk management
b. Strategies for natural resources management (water, soil, …)
c. Land use planning
d. Facilities supply
e. Urban (and/or rural) development
f. Other

4. Do you consider peri-urban dynamics and in planning? Have you specific policies and strategies for peri-urban areas? What? Are there examples of planning in PU areas? What about results?

5. What do you do for sustainable resource management? What about soil, water?

6. What about donors: names, role/tasks, programs and approaches?

7. What about fiscal system in PU areas?

8. What about land tenure in PU areas?

9. What about facilities supply in PU areas?

a. Transport system
b. Electricity
c. Water
d. Waste management
e. Education
f. Health
g. Other

10. Have you observes changes in PU (U and R) development during the last year? What?

11. According to your experiences what are the main relations between PU-U and PU-R areas? Have those changed in the last period?

a. Flows of goods
b. People
c. Other

12. What is estimated amount of solid waste generated per day?

a. 1 bucket
b. 1–3 buckets
c. More than 3 buckets

13. Do you know about networks and connections, relations of trust and mutual support, formal and informal groups in PU areas? How do they contribute to decision-making?

14. Have you observed environmental changes during the last year?

a. Water availability
b. Soil fertility
c. Soil aridity
d. Humidity
e. Rain pattern
f. Other

15. Have you evaluated the impact of these changes in PU areas?

16. Do you implement any strategies to cope with these change and top reduce PU? vulnerability

17. Have you observed autonomous adaptation to these changes? For example changes in PU activities.

a. Change in crop system
b. Change in livestock
c. Change on house structure

18. Do you have data or baseline study on PU dwellers composition and livelihoods?

a. Employment
b. Population
c. Type and house dimension
d. Other

19. What are the reasons of these environmental problems?

a. Climatic variation
b. Human land use
c. Inadequate environmental management from institutions
d. Other

20. Notes

Annex II
Glossary

Adaptation (IPCC 2014)
The process of adjustment to actual or expected climate and its effects. In human systems, adaptation seeks to moderate harm or exploit beneficial opportunities. In natural systems, human intervention may facilitate adjustment to expected climate and its effects.

> ***Incremental adaptation*** Adaptation actions where the central aim is to maintain the essence and integrity of a system or process at a given scale.
> ***Transformational adaptation*** Adaptation that changes the fundamental attributes of a system in response to climate and its effects.

Anticipatory adaptation—Adaptation that takes place before impacts of *climate change* are observed. Also referred to as proactive adaptation (IPCC 2007).

Autonomous adaptation—Adaptation that does not constitute a conscious response to climatic stimuli but is triggered by ecological changes in natural systems and by market or *welfare* changes in *human systems*. Also referred to as spontaneous adaptation (IPCC 2007).

Planned adaptation—Adaptation that is the result of a deliberate policy decision, based on an awareness that conditions have changed or are about to change and that action is required to return to, maintain or achieve a desired state (IPCC 2007).

Adaptation assessment (IPCC 2007)
The practice of identifying options to adapt to *climate change* and evaluating them in terms of criteria such as availability, benefits, costs, effectiveness, efficiency and feasibility.

Adaptation benefits (IPCC 2007)
The avoided damage costs or the accrued benefits following the adoption and implementation of *adaptation* measures.

Adaptive capacity (IPCC 2014)
The ability of systems, institutions, humans, and other organisms to adjust to potential damage, to take advantage of opportunities or to respond to consequences.

L. Ricci, *Reinterpreting Sub-Saharan Cities through the Concept of Adaptive Capacity*, SpringerBriefs in Environment, Security, Development and Peace 26, DOI 10.1007/978-3-319-27126-2

Adaptive management (IPCC 2014)
A process of iteratively planning, implementing, and modifying strategies for managing resources in the face of uncertainty and change. Adaptive management involves adjusting approaches in response to observations of their effect and changes in the system brought on by resulting feedback effects and other variables.

Climate sensitivity (IPCC 2014)
In IPCC reports, equilibrium climate sensitivity (units: °C) refers to the equilibrium (steady state) change in the annual global mean surface temperature following a doubling of the atmospheric equivalent carbon dioxide concentration. Due to computational constraints, the equilibrium climate sensitivity in a climate model is sometimes estimated by running an atmospheric general circulation model coupled to a mixed-layer ocean model, because equilibrium climate sensitivity is largely determined by atmospheric processes. Efficient models can be run to equilibrium with a dynamic ocean. The climate sensitivity parameter (units: °C $(W\ m^{-2})^{-1}$) refers to the equilibrium change in the annual global mean surface temperature following a unit change in radiative forcing.

Contextual vulnerability (Starting-point vulnerability) (IPCC 2014)
A present inability to cope with external pressures or changes, such as changing climate conditions. Contextual vulnerability is a characteristic of social and ecological systems generated by multiple factors and processes (O'Brien et al. 2007).

Outcome vulnerability (End-point vulnerability) (IPCC 2014)
Vulnerability as the end point of a sequence of analyses beginning with projections of future emission trends, moving on to the development of climate scenarios, and concluding with biophysical impact studies and the identification of adaptive options. Any residual consequences that remain after adaptation has taken place define the levels of vulnerability (Kelly and Adger 2000; O'Brien et al. 2007).

Resilience (IPCC 2014)
The capacity of a social-ecological system to cope with a hazardous event or disturbance, responding or reorganizing in ways that maintain its essential function, identity, and structure, while also maintaining the capacity for adaptation, learning, and transformation (Arctic Council 2013).

Sensitivity (IPCC 2014)
The degree to which a system or species is affected, either adversely or beneficially, by climate variability or change. The effect may be direct (e.g. a change in crop yield in response to a change in the mean, range, or variability of temperature) or indirect (e.g. damages caused by an increase in the frequency of coastal flooding due to sea-level rise).

Vulnerability (IPCC 2007)
Vulnerability is the degree to which a system is susceptible to and unable to cope with adverse effects of *climate change*, including *climate variability* and extremes. Vulnerability is a function of the character, magnitude, and rate of climate change and variation to which a system is exposed, its *sensitivity*, and its adaptive capacity.

Vulnerability (IPCC 2014)
The propensity or predisposition to be adversely affected. Vulnerability encompasses a variety of concepts including sensitivity or susceptibility to harm and lack of capacity to cope and adapt. See also Contextual vulnerability and Outcome vulnerability.

Source: Glossary of Terms used in the IPCC Fourth Assessment Report, WG II; at: http://www.ipcc.ch/pdf/glossary/ar4-wg2.pdf and Annex II Glossary of the IPCC Fifth Assessment Report, Working Group II (2014) http://www.ipcc.ch/report/ar5/wg2/.

Adaptation
Initiatives and measures to reduce the vulnerability of natural and human systems against actual or expected *climate change* effects. Various types of adaptation exist, e.g. *anticipatory* and *reactive*, *private* and *public*, and *autonomous* and *planned*. Examples are raising river or coastal dikes, the substitution of more temperature-shock resistant plants for sensitive ones, etc.

Adaptive capacity
The whole of capabilities, resources and institutions of a country or *region* to implement effective *adaptation* measures.

Climate sensitivity
In IPCC reports, equilibrium climate sensitivity refers to the equilibrium change in the annual mean global surface temperature following a doubling of the atmospheric equivalent carbon dioxide concentration. Due to computational constraints, the equilibrium climate sensitivity in a climate model is usually estimated by running an atmospheric general circulation model coupled to a mixed-layer ocean model, because equilibrium climate sensitivity is largely determined by atmospheric processes. Efficient models can be run to equilibrium with a dynamic ocean. The transient climate response is the change in the global surface temperature, averaged over a 20-year period, centred at the time of atmospheric carbon dioxide doubling, that is, at year 70 in a 1% per year compound carbon dioxide increase experiment with a global coupled climate model. It is a measure of the strength and rapidity of the surface temperature response to greenhouse gas forcing.

Resilience (see above)
Sensitivity (see above)
Vulnerability (see above)

Source: Glossary of IPCC (2007), Fourth Assessment Report, Synthesis Report, http://www.ipcc.ch/pdf/assessment-report/ar4/syr/ar4_syr_appendix.pdf.

Other definitions of Adaptation

A process by which strategies to moderate, cope with and take advantage of the consequences of climatic events are enhanced, developed and implemented (UNDP 2005).

Adaptation to climate change is the process through which people reduce the adverse effects of climate on their health and well-being, and take advantage of the opportunities that their climatic environment provides (Burton 1992, cited in Smit et al. 2000).

Adaptation involves adjustments to enhance the viability of social and economic activities and to reduce their vulnerability to climate, including its current variability and extreme events as well as longer term climate change (Smit 1993, cited in Smit et al. 2000).

The term adaptation means any adjustment, whether passive, reactive or anticipatory, that is proposed as a means for ameliorating the anticipated adverse consequences associated with climate change (Stakhiv 1993, cited in Smit et al. 2000).

Adaptation to climate change includes all adjustments in behavior or economic structure that reduces the vulnerability of society to changes in the climate system (Smith et al. 1996, cited in Smit et al. 2000)

Source: Smith, Stephen C., Arun Malik and Xin Qin "Autonomous Adaptation to Climate Change: A Literature Review". IIEP Working Paper 2010-27.

Annex III
Land Tenure System in Tanzania

Different tenure systems	
Statutory	Customary
Granted right of occupancy The governments grants its citizens renewable rights of occupancy on land that has been surveyed of up to 99 years at a premium and revisable annual land rent. To be valid, the right has to be registered under the Land Registration Ordinance Chapter 334.	*Customary* Is acquired by virtue of being a member of a community and is based on traditional acceptance. The system has no formal documents and no land transfer takes place without the blessings of the clan/community members
Occupancy under Letter of Offer Once a citizen is issued with and accepted a letter of offer, he/she can register the duly signed and sealed letter of offer under Registration of documents Ordinance Chapter 117 and becomes a valid document that creates notice of ownership	*Quasi-customary o neo-customary tenure* As the name suggests, the influence of the clan/community in land transfer is, among other things, diminished. While local leaders and adjoining landowners are consulted when the need to transfer land arises, the right to sell lies mainly with the individual right holder. Customary and quasicustomary forms of tenure are commonly found in peri-urban unplanned areas of the city of Dar es Salaam (Kombe 1995; Kironde 2005)
Derivative right Under the Land Act (1999) the government offers a 'residential licence', which is a right derivative of a granted right of occupancy. According to the Act, a residential licence is a right conferred upon the licensee to occupy land in non-hazardous land, land reserved for public utilities and surveyed land for a term not less than six months and not more than two years. The term can, however, be renewed for the same period. Like occupancy under letter of offer, the residential residence is issued under Registration of documents Ordinance Chapter 117	*Informal tenure* Here, land transfer is not guided by customary or quasi-customary norms and rules. It can take place between any land seeker and the person who owns the land and the system has devised its own informal ways of protecting the buyer and authenticating ownership (Kombe 1995)

Source Sheuya, S.A., 2010, Informal settlements and finance in Dar es Salaam, Tanzania. Nairobi: UN-HABITAT; Kombe, W.J. 2010, Land Conflicts in Dar es Salaam: Who gains? Who loses?, Crisis States. Working Paper 82 (series 2) London: Crisis States Research Centre

L. Ricci, *Reinterpreting Sub-Saharan Cities through the Concept of Adaptive Capacity*, SpringerBriefs in Environment, Security, Development and Peace 26, DOI 10.1007/978-3-319-27126-2

Annex IV
Adaptation Opportunities in Cities

Sector	Adaptation option/strategy	Underlying policy framework	Key constraints (−) or opportunities (+) to implementation
Water	Expanded rainwater harvesting water storage and conservation techniques; water re-use; desalination; water-use and irrigation efficiency	National water policies and integrated water resources management; water-related hazards management	(−) Financial, human, resources and physical barriers
			(+) Integrated water resources management; synergies with other sectors
Infrastructure and settlement (including coastal zones)	Relocation; seawalls and storm surge barriers; dune reinforcement; land acquisition and creation of marshlands/wetlands as buffer against sea-level rise and flooding; protection of existing natural barriers	Standards and regulations that integrate climate change considerations into design; land use policies; building codes; insurance	(−) Financial and technological barriers
			(+) Availability of relocation space, integrated policies and managements, synergies with sustainable development goals
Human health	Heat-health action plans, emergency medical services, improved climate-sensitive disease surveillance and control, safe water and improved sanitation	Public health policies that recognize climate risk; strengthened health services; regional and international cooperation	(−) Limits to human tolerance (vulnerable groups)
			(−) Knowledge limitations
			(−) Financial capacity
			(+) Upgraded health services
			(+) Improved quality of life

(continued)

L. Ricci, *Reinterpreting Sub-Saharan Cities through the Concept of Adaptive Capacity*, SpringerBriefs in Environment, Security, Development and Peace 26, DOI 10.1007/978-3-319-27126-2

Sector	Adaptation option/strategy	Underlying policy framework	Key constraints (−) or opportunities (+) to implementation
Tourism	Diversification of tourism attractions and revenues, shifting ski slopes to higher altitudes and glaciers	Integrated planning (e.g., carrying capacity; linkages with other sectors); financial incentives, e.g., subsidies and tax credits	(+) Appeal/marketing of new attractions (−) Financial and logistical challenges (−) Potential adverse impact on other sectors (e.g., artificial snowmaking may increase energy use) (+) Revenues from new attractions (+) Involvement of wider group of stakeholders
Transport	Realignment/relocation; design standards and planning for roads, rail, and other infrastructure to cope with warming and drainage	Integrating climate change considerations into national transport policy; investment in R&D for special situations, (e.g., permafrost areas)	(−) Financial and technological barriers (−) Availability of less vulnerable routes (+) Improved technologies (+) Integration with key sectors (e.g., energy)
Energy	Strengthening of overhead transmission and distribution infrastructure, underground cabling for utilities, energy efficiency, use of renewable resources, reduced dependence on single sources of energy	National energy policies, regulations, and fiscal and financial incentives to encourage use of alternative sources; incorporating climate change in design standards	(+) Access to table alternatives (−) Financial and technological barriers (−) Acceptance of new technologies (+) Stimulation of new technologies (+) Use of local resources

Source Romero-Lancao 2010 tratto da IPCC (2007), World Bank (2008)

Sapienza University of Rome

Sapienza University of Rome, officially *Sapienza – Università di Roma*, also called simply *Sapienza* formerly known as *Università degli studi di Roma "La Sapienza"*, is a collegiate research university. It is the largest European university by enrollments and one of the oldest, founded in 1303. Sapienza has been a high performer among the largest internationally ranked universities. Since its foundation more than 700 years ago, Sapienza has played an important role in Italian history and has been directly involved in key changes and developments in society, economics, and politics. It has contributed to the development of Italian and European science and culture in all areas of knowledge, and educated numerous notable alumni, including many Nobel laureates, presidents of the European Parliament, heads of several nations, and significant scientists and astronauts.

Following its 2011 reform, Sapienza University of Rome now has eleven faculties and 65 departments. Today, with 140,000 students and 8,000 academic, technical and administrative staff, Sapienza is the largest university in Italy. The university has significant research programmes in the fields of engineering, natural sciences, biomedical sciences, and humanities. It offers 10 Masters Programmes taught entirely in English.

Sapienza offers a vast array of courses including degree programmes, Ph.D. courses, professional courses, and Specialization Schools in many disciplines, run by 63 Departments and 11 Faculties. Sapienza carries out outstanding scientific research in most disciplines, achieving impressive results on the national and international level. Sapienza also enhances research by offering opportunities to academia on a global scale. Thanks to a special programme for visiting professors, many foreign researchers and lecturers periodically visit the University and contribute greatly to the quality of education and research programmes. Some 8,000

L. Ricci, *Reinterpreting Sub-Saharan Cities through the Concept of Adaptive Capacity*, SpringerBriefs in Environment, Security, Development and Peace 26, DOI 10.1007/978-3-319-27126-2

foreign students are regularly enrolled at Sapienza and there are over 1,100 incoming and outgoing exchange students each year with the support of several mobility programmes.

Department of Civil, Constructional and Environmental Engineering (DICEA)

The *Department of Civil, Building, and Environmental Engineering* (DICEA) at the Faculty of Engineering of Sapienza University of Rome was established on 1 July 2010, and merged the cultural traditions of the Departments of Architecture, Urban Planning for Engineering, and Hydraulics, Transport, and Roads. Its teaching and research focuses on a wide variety of disciplines in architecture and urban planning, civil engineering, and environmental engineering.

DICEA contributes to several undergraduate and graduate programs in the School of Civil and Industrial Engineering and offers three Ph.D. programmes: Infrastructure and Transport, Hydraulics and Environmental Engineering, and Architecture and Urban Planning Engineering. DICEA also contributes to post-graduate training through fellowships and research grants, and offers two post-graduate degrees, in Railway Infrastructure and Systems Engineering and in Management and Maintenance of Ecological Plants.

DICEA is actively engaged in research in frontier areas of its various fields of expertise, which places it in a top position on the national and international scene in both basic and applied research. Due to its close collaboration with public authorities and institutions and with business and services companies, DICEA has acquired a strong ability to develop testing applications and field studies. The experimental approaches used are largely multi- and interdisciplinary, as increasingly required by the complexity of the issues addressed and the pursuit of methodological innovation.

The main research areas of DICEA are:

- Technical Architecture
- Hydraulic Infrastructure
- Geodesy and Geomatics
- Geophysics
- Geology and Geotechnology
- Hydraulics
- Environmental Engineering
- Roads
  - Transportation
  - Architectural and Urban Design
  - Urban Planning

The extensive research produced by DICEA is evinced by the number of scientific publications it has produced, the many national and international conferences and exhibitions it has organized, its active involvement in a number of national and international research networks, the national and international awards obtained, and the creation of patents.

DICEA's research is characterized by a strong internationalization and operates through a variety of activities based on diverse funding sources, both public and private, national and international: government research programmes (PRIN, FIRB, etc.) and research institution programmes (CNR, etc.); grants and tenders from the EU and other international organizations; regional and local government authorities; agencies and public institutions (land reclamation consortia, river basin authorities, etc.); public services companies (in the fields of transport, energy, waste, water, infrastructure, mobility, etc.); University funding, etc. As a natural extension of these research activities, a number of university spin-offs have been initiated, which also offer opportunities to support the activities of young researchers.

About the Author

Liana Ricci (Italy) is an Environmental Engineer and has a Ph.D. in Urban Planning. She is currently a post-doc fellow in the Department of Civil, Building and Environmental Engineering, at Sapienza University of Rome, Italy. Her major fields of specialization are urban and environmental planning in sub-Saharan cities, adaptive capacity assessment, climate change adaptation mainstreaming into urban policy and plans, and urban policy mobility. Liana has conducted research in methodology development for data collection and analysis with respect to spontaneous practices for climate change adaptation in urban areas of Least Developed Countries, Local Agenda 21, and in Environmental Planning and Management in peri-urban areas.

Her work has been presented at several national and international conferences and workshops (INURA; AESOP; ICSS 2010; European Ph.D. Network on International Climate Policy 2010; Ph.D. Workshop on Sustainable Development, Columbia University 2011, UICCA conference, Turin 2011 and 2013, UGRG 2014). In 2013 she was a blue book trainee at the European Commission, DG Clima, working as a Policy Assistant in the team for Mainstreaming of Climate Action into Cohesion Policy for the programming period 2014–2020 in the macroregions, and supported the development of the Knowledge Gap Strategy for Climate Action (Horizon 2020). During the same year she also participated in the UN Habitat Internship Programme on Urban Planning to stimulate and contribute to research related to the *Cities and Climate Chance Initiative* (CCCI) and participated in the CCCI Technical Support Team.

Major Publications

Ricci, L., Sanou, B., Baguian, H. (2015) Climate risks in West Africa: Bobo-Dioulasso local actors' participatory risks management framework, Current Opinion in Environmental Sustainability, Vol. 13, April 2015, Pages 42–48, ISSN 1877-3435, http://dx.doi.org/10.1016/j.cosust.2015.01.004 (link is external)

L. Ricci, *Reinterpreting Sub-Saharan Cities through the Concept of Adaptive Capacity*, SpringerBriefs in Environment, Security, Development and Peace 26, DOI 10.1007/978-3-319-27126-2

Macchi, S., Ricci, L. (2014) Mainstreaming Adaptation into Urban Development and Environmental Management Planning: a literature review and lesson from Tanzania. Macchi, S., Tiepolo, M. (eds), Climate Change Vulnerability in Southern African Cities. Springer Climate. ISBN 978-3319006710, pp. 109–124 http://link.springer.com/chapter/10.1007%2F978-3-319-00672-7_7 (link is external)

Ricci, L. (2014) Linking adaptive capacity and peri-urban features. Macchi, S., Tiepolo, M. (eds), Climate Change Vulnerability in Southern African Cities. Springer Climate. ISBN 978-3319006710, pp. 89–107 http://link.springer.com/chapter/10.1007%2F978-3-319-00672-7_6 (link is external)

Congedo, L., Macchi S., Ricci, L., Faldi G. (2013) Urban sprawl e adattamento al cambiamento climatico: il caso di Dar es Salaam. Planum. The Journal of Urbanism, no. 27, vol. 2./2013. ISSN 1723-0993

Kehew, R., Rollo, C., Mthobeli, K., Callejas, A., Alber, G., & Ricci, L. (2013). Formulating and implementing climate change laws and policies in the Philippines, Mexico (Chiapas) and South Africa: A local government perspective. Local Environment: International Journal of Justice and Sustainability, 18(6), 723–737. ISSN 1354-9839

Address: Dr. Liana Ricci, SAPIENZA University of Rome, DICEA, Architettura e Urbanistica, Edificio A—Piano 2°, Area Architettura e Urbanistica, Via Eudossiana 18, 0184 Roma, Italy
Email: liana.ricci@uniroma1.it
Website: http://www.researchgate.net/profile/Liana_Ricci2 and http://www.dicea.uniroma1.it/it/users/lianaricci.

About this Book

This book explores whether and how a reinterpretation of sub-Saharan cities, through the concept of adaptive capacity, could contribute to a new understanding of the contemporary city. The research improves knowledge of urban and environmental planning and the dynamics of development and environmental management in peri-urban areas of Dar es Salaam. This knowledge highlights the limits of certain common generalizations on the character of peri-urban areas. Moreover, the research assesses the strengths and weakness of tools and methods for investigating adaptive capacity and environmental management in the city of Dar es Salaam. Finally, it highlights controversial issues and possible research paths related to the relationship between adaptive capacity and urban and environmental planning.

- This volume includes original knowledge on autonomous adaptation and environmental management in Dar es Salaam;
- It provides an original integration of adaptation and urban planning concepts in a post-colonial framework;
- It offers a new perspective on the coupled human-environment system and its relevance for adaptation.

Contents:

More on this book at: http://www.afes-press-books.de/html/SpringerBriefs_ESDP_21.htm.

L. Ricci, *Reinterpreting Sub-Saharan Cities through the Concept of Adaptive Capacity*, SpringerBriefs in Environment, Security, Development and Peace 26, DOI 10.1007/978-3-319-27126-2

Printed in the United States
By Bookmasters